The return of
HALLEY'S COMET

PATRICK MOORE
& JOHN MASON

Patrick Stephens, Cambridge

0 10'

First published in 1984

ISBN 0-85059-667-X

Photoset in 10 on 11 pt Garamond by Manuset Ltd, Baldock, Herts. Printed in Great Britain on 115 gsm Fineblade coated cartridge, and bound, by Anchor Brendon Limited, Tiptree, Colchester, for the publishers, Patrick Stephens Limited, Bar Hill, Cambridge, CB3 8EL, England.

CONTENTS

PREFACE

One of the most famous of all celestial visitors is, without doubt, Halley's Comet. It has been seen every 76 years or so for many centuries; records of it go back to long before the Christian era, and when it appears it never fails to cause widespread interest—popular as well as scientific.

As we write these words (September 1983) Halley's Comet is on its way back. As yet it is very dim, but it will brighten steadily until it becomes a naked-eye object in late 1985. It is of great importance to astronomers and, though conditions at this return will be far from ideal, they will at least provide many people with their only chance of seeing this, the most celebrated of all comets.

Therefore, we feel that it is appropriate to produce a book which may help those who are anxious to make the most of an opportunity which will not recur until 2061 AD. We hope that what we have written will be of interest. If not, then the fault will be ours, and not that of Halley's Comet!

Patrick Moore, Selsey
John Mason, Arundel

September 1983

CHAPTER 1

COMETS: ECCENTRIC VISITORS FROM SPACE

On October 16 1982, two astronomers were busy at the Palomar Observatory in California. This is one of the most famous observatories in the world, and contains the 5.1-metre or 200-inch Hale reflector, which was for many years the largest telescope in existence and is even now surpassed by only one other. Generally, the Hale reflector is used for studying faint stars and star systems which are beyond the range of smaller telescopes. On this occasion, however, the two astronomers—G.E. Danielson and David C. Jewitt—were concentrating upon something much nearer home. They were hunting for Halley's Comet.

Of all comets, Halley's is much the most famous; it comes back every 76 years on average, and is the only periodical comet which can become conspicuous as seen with the naked eye. Its closest approach to the Sun, or perihelion, was not due until February 1986, so that during 1982 it was bound to be excessively faint, but Danielson and Jewitt had high hopes; they were using the world's second largest telescope together with an electronic device known as a CCD, which is much more sensitive than any photographic plate.

They checked their results. There, almost exactly in the expected position, was a tiny speck. Before long they were certain of their success, and by October 20 they were confident enough to announce it to the world. By then, the comet had also been picked up by M. Belton and H. Butcher, using a CCD on the 4-metre reflector at the Kitt Peak Observatory in Arizona.

The comet was indeed faint; magnitude 24.2—but it was unquestionably present, and its recovery, for the first time since it faded from view in 1911, caused tremendous interest, not entirely confined to the astronomical fraternity. The early observations meant that the comet could be kept under constant scrutiny as it drew inward toward the Sun, brightening slowly at first and then more and more rapidly as it entered the inner regions of the solar system. And Halley's Comet has much to tell us, particularly inasmuch as there is a great deal about comets about which our knowledge is still surprisingly small.

Of course, comets have been recorded since very early times. The ancient Chinese and Japanese were very interested in them, and studied their movements against the starry background; their observations have been found to be very useful in modern times. They had no idea of the nature of a comet—and even in Europe, up to the Middle Ages, comets were believed to be luminous bodies floating in the Earth's upper air. It was not until 1577 that the great Danish astronomer Tycho Brahe made careful observations of the comet of that year, and showed that it must be several times as far away as the Moon. Also, all brilliant comets with the exception of Halley's are unpredictable. They return to the neighbourhood of the Sun only once in many centuries, thousands of years or even millions of years, so that we never know when or where to expect them.

Nowadays, one of the most common questions asked of the astronomer by the layman runs along these lines: 'Last night I saw a bright point of light flash very quickly across the sky. Can it have been a comet?' The invariable answer is 'no'. The object was almost certainly a meteor or shooting-star, unless it could have been a quickly-moving artificial satellite orbiting several hundred kilometres above the Earth's surface. A comet is a long way away—far beyond the top of the air—and so it does not move perceptibly. It has to be watched for some time before its movement against the sky can be noticed.

The view of a bright comet in the dawn or dusk sky is an impressive spectacle—though it must be admitted that really brilliant visitors have been much less common in our own century than they were during the last, and nobody now living can remember a comet as magnificent as, for instance, those of 1811, 1843 or 1858. The last comet easily visible in daylight for an appreciable period was that of 1910, known as the Daylight Comet (and, incidentally, it was far brighter than Halley's, which followed later in the same year). Still, there have been several recent naked-eye comets, such as those of 1957 (Arend-Roland), 1970 (Bennett), 1976 (West), and 1983 (IRAS-Araki-Alcock), and it is at first sight rather difficult to believe that the entire cometary phenomenon originates in a tiny central nucleus, probably no more than a few kilometres in diameter, composed of what astronomers call 'dirty ice'—that is to say, ordinary ice impregnated with small amounts of other chemical elements. The interaction of this dirty snowball, as it is termed, with the Sun's radiation, and in particular the solar wind, results in the production of all the typical cometary characteristics, especially the dust and gas tails for which comets are well known. The study of comets encompasses many other subject areas of astronomy. They are believed to be the most primitive material within our solar system, little altered since the formation of that system some 4½ thousand million years ago. The study of comets may give us an insight into the early history of the solar system, as they are thought to have formed from the primeval solar nebula at approximately the same time as the planets and other solar system bodies. Cometary material has probably changed little since their formation, and this is one reason why space probes are being despatched for observations at close hand, to determine the physical and chemical characteristics of the nucleus, and study the whole range of cometary phenomena. Furthermore, by monitoring the appearances of comets and their tails, we learn much about their interaction with the various emanations from the Sun, including radiation, highly energetic atomic particles, and solar magnetic field. The tremendous dependence of cometary phenomena on the Sun means that they are all of a transient nature, being directly related to the distance of the comet from the Sun and the solar activity, at any time.

Additional information on cometary material may be obtained by means of the particles of solid debris which they distribute throughout interplanetary space. These particles become visible when they are caught up in the Earth's gravitational field, are pulled into the upper layers of the Earth's atmosphere and are quickly incinerated, producing the brief streak of light which we call a meteor. The solid particles present in the interplanetary medium may also scatter or absorb sunlight, to produce the zodiacal light and the *Gegenschein*. From the tropics, the zodiacal light is a striking feature for a visual observer, appearing as a cone of light in the west, after sunset, and in the east, before sunrise. It is brightest nearer the direction of the Sun and through the constellations of the zodiac.

In general, a comet will appear in the inner solar system as a slow-moving, rather ill-defined hazy patch of light, usually with a central condensation. The first observations of comets are, naturally, their discovery. The brighter ones are usually first detected by amateur astronomers, using wide-field telescopes or even binoculars. The fainter comets are more often discovered by professional astronomers, on photographic plates taken for some other purpose. All cometary discoveries are communicated to the Bureau for Astronomical Telegrams, at the Smithsonian Astrophysical Observatory, Cambridge, Massachusetts, and are then published in the International Astronomical Union Circulars. Comets are normally named after their discoverers, and, currently, up to three independent co-discoverers are permitted; Comets Honda-Mrkós-Pajdusáková and Tuttle-Giacobini-Kresák are two such examples. This rule is understandable if one imagines the confusion which would occur if some bright comet was discovered almost simultaneously by several dozen different people world-wide! Very occasionally, a comet has been officially named after the astronomer who first computed its orbit; two examples are Comet Halley and Comet Encke.

In addition to being named, comets also receive two different types of designation. The first comet discovered in 1983 would be labelled 1983a, the second 1983b and so on, for other discoveries throughout the year. Later, when the precise orbit of the comet around the Sun has been calculated, a second form of designation is issued. In each year, comets are given a Roman numeral, in order of their times of closest approach to the Sun, the point known as perihelion passage. For example, the second comet to pass perihelion in 1983 would be 1983 II and the fourth, 1983 IV and so on. This can cause some confusion for newcomers to the subject, if a comet discovered in one year passes perihelion in a different year. For example, Bennett's Comet, which was a spectacular object in the spring of 1970 (see Fig 1.1), first received the designation 1969i (the ninth comet to be discovered in 1969), but was later known as 1970 II, being the second comet to pass perihelion in that year. In the case where a newly-discovered comet is recognised as a known periodic comet which has been seen before, the comet continues to bear the name of its original discoverer, and the more recent rediscoverer of the object merely gets a pat on the back. Finally, known periodic comets having a period of less than 200 years, are preceded by P/. For example, Comet P/Halley and Comet P/Grigg-Mellish. In the ten years between 1964 and 1973 inclusive, 108 comet discoveries were made, but 56 of these were recoveries of known periodic comets.

Once a comet is discovered, a minimum of three precise measurements of its position in the sky, relative to the background stars, are sufficient to compute an orbit, but in practice, accurate orbits are only derived from many precise observations. Six parameters are necessary to define a cometary orbit. These parameters define the size and shape of the orbit, and the orientation of the orbital plane in space, relative to the plane of the Earth's orbit, and also the time of perihelion passage of the comet. Cometary orbits are often elliptical in form, similar to the orbits of the planets but, in general, those of comets are much more elongated and have a wide range of angles of inclination to the plane of the Earth's orbit. It was the great German mathematician, Johannes Kepler, who first showed, between 1609 and 1618, that the planets moved around the Sun in ellipses, not circles. Within any ellipse are two points known as the

Fig 1.2 Orbital eccentricities

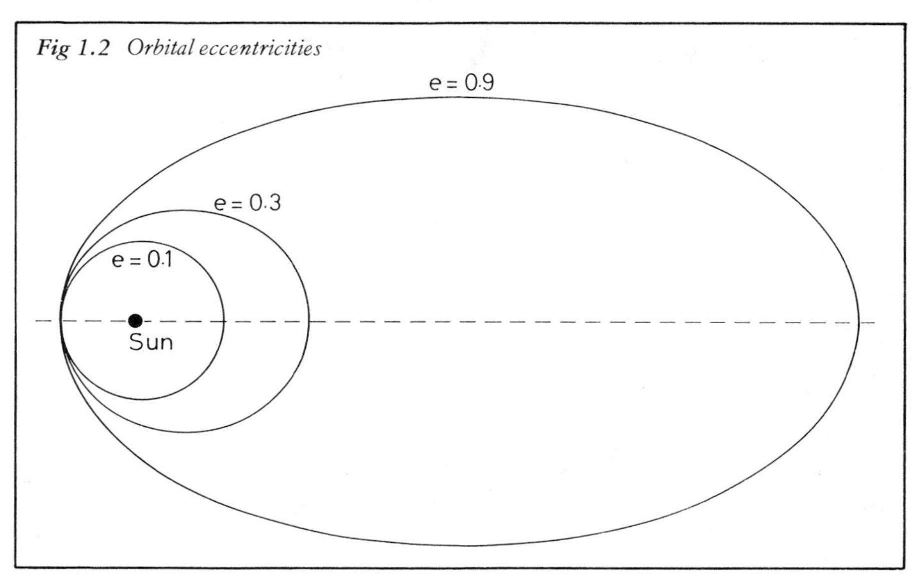

Left *Fig 1.1 Bennett's Comet in the spring of 1970.*

foci. For the planets, the Sun lies at one focus of the orbit, while the other focus is empty. The ratio of the distance between the two foci, to the whole length of the elliptical orbit, is a measure of something called the orbital eccentricity, denoted by the letter 'e'. If the distance between the foci is zero, you have a perfect circle with no eccentricity at all. The planetary orbits are ellipses which are very close to circles. The most eccentric planetary orbit is that of Pluto, for which e = 0.256. Because the eccentricity is a ratio, it must also be less than unity for the orbit to be closed. The more the eccentricity approaches the value 1, the longer and narrower the ellipse becomes. The majority of cometary orbits are of this form. Three elliptical orbits having eccentricities of 0.1, 0.3 and 0.9 are shown in Fig 1.2 for comparison. When the eccentricity is e = 1, the orbit is no longer a closed loop and we have an open curve of the kind known as a parabola. Obviously, a body moving on a parabolic orbit would only pass by the Sun once. Thereafter, it would continue moving outwards indefinitely. Some orbits are even more open than the parabola, and these are said to be hyperbolic. A hyperbola has an eccentricity of greater than 1.

In practice, many precise observations are required to evaluate accurately the orbital eccentricity of a comet. Unfortunately, more than half of the comets seen to date have not been observed well enough to distinguish their orbits from parabolas. To facilitate the computation of the remaining five orbital elements, the eccentricity 'e' is often assumed to be unity, (ie, 1.00). However, it is quite likely that 'e' could be either 0.999, or 1.001, and so it follows that, saying a particular comet has a parabolic orbit (e = 1) is really a statement of ignorance. It is just possible the orbit is really hyperbolic, but far more likely that it is an ellipse of very long period, possibly millions of years. Unfortunately, we can only observe comets when they are relatively close to the Sun, and they are usually observable over only a tiny fraction of their complete orbit. You can see from Fig 1.3 that if we only observe the comet between points A and B in the diagram, the three distinct orbital forms; ellipse, parabola and hyperbola, all look very much the same.

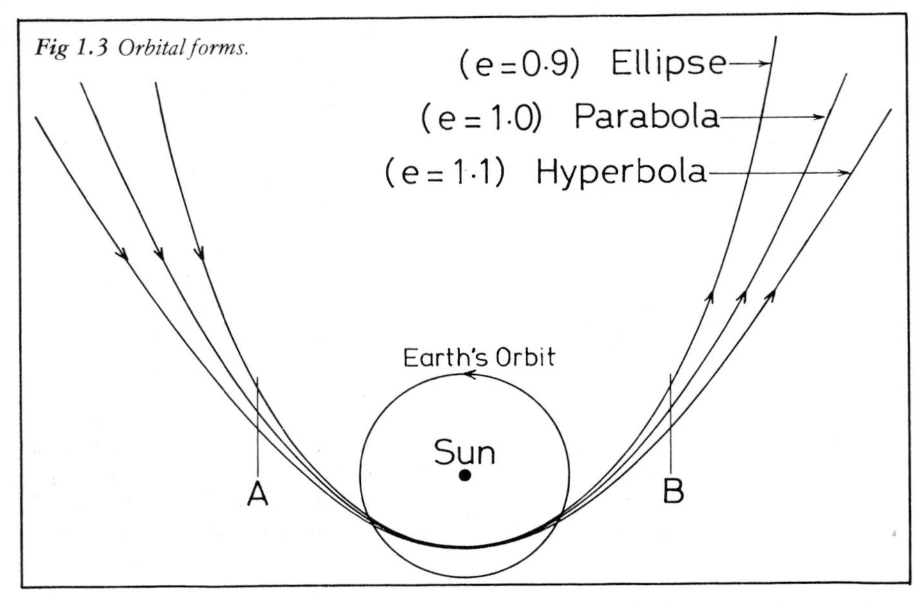

Fig 1.3 Orbital forms.

Comets are classified according to their derived orbits. A short period comet has a period of less than 200 years, and will have an elliptical orbit. The catalogue produced by Brian Marsden of the Smithsonian Astrophysical Observatory in 1979, contains the orbital elements for 1,027 cometary apparitions, observed up to the end of 1978, but it refers to only 658 individual comets, the remainder representing the various returns of the periodic comets. Of the 658 different comets, only 113 are classified as 'periodic' with a maximum period of 155 years. Of this number 72 comets have been observed at two or more perihelion passages, and hence have quite well-established orbits. The 'average' short period comet has a perihelion distance of 1.5 Astronomical Units (1AU = 149,597,870 km), an orbital inclination of 13° and a return period of 7 years. A large number of this class of comets have their aphelion distances (farthest points from the Sun) clustered at about 5.3 AU from the Sun, very close to the orbit of the planet Jupiter, and are consequently known as the Jupiter 'family' of comets. The gas-giant Jupiter, nearly 320 times as massive as our Earth, has a profound gravitational influence upon these comets. The Jupiter family usually includes only those comets with periods of less than 13 years; the period of Jupiter itself is 11.86 years. In fact, as will be discussed later, Jupiter is largely responsible for the very existence of the shortest period comets.

So far we have only discussed comets with periods of less than 200 years. A long period comet is defined as one with a period greater than 200 years, and 545 of these are listed in Marsden's 1979 Catalogue. Of this number 285 appear to have parabolic orbits, 162 elliptical orbits and the remaining 98 hyperbolic orbits, the greatest orbital eccentricity in the catalogue being 1.006. It was shown as long ago as 1914, by Strömgren, that when these apparently hyperbolic orbits in the inner solar system are traced backwards in time, to beyond the orbit of Neptune, they turn out to have been elliptical originally. It is now believed that there is a probability of only 1 in 10,000 that any of the slightly hyperbolic comets emanated from interstellar space. Any comet moving from interstellar space should enter the solar system with a much greater velocity than we actually observe. It is found that the comets with the greatest orbital eccentricity all tend to have large perihelion distances.

All three forms of cometary orbit, have as one of their six orbital elements the semi-major axis of the conic section denoted by the letter 'a'. For an ellipse, 'a' is greater than zero but not infinite; for a parabola 'a' is infinite, and for a hyperbola 'a' is less than zero (ie, 'a' is negative). The period of a comet in years, for all values of 'a' greater than zero (measured in Astronomical Units) is given by:

$$P \text{ (in years)} = a^{3/2} \text{ (for } a>0)$$

A remarkable and exciting result is obtained if one examines the distribution in the values of the reciprocal of the semi-major axis 1/a for the 86 comets with nearly parabolic orbits which have been most accurately determined, corrected for the perturbations of the planets. The graph is shown in Fig 1.4 and the peak at a value of $1/a = 4\times10^{-5}(AU)^{-1}$ is striking. The orbital period corresponding to this value of 1/a is $(25,000)^{3/2}$, or roughly 4 million years. A comet having this period would have an aphelion distance of about 50,000 AU, which is nearly 0.8 light years, or almost $\frac{1}{5}$ of the distance to the next nearest star from the Sun. In 1950, Jan Oort, of the University of Leiden, gave a classic explanation of this peak. He postulated that comets must be stored in a vast cloud having the Sun at the centre, extending out to roughly 50,000 AU. The comets within this cloud would have perihelion distances well beyond the very outer reaches of our solar system. This huge 'store-house' of extremely long period comets became known as Oort's Cloud.

Our Sun is a perfectly ordinary star and is just one member of the great assembly of some 100 thousand million stars which make up our Milky Way Galaxy. If the Sun were alone is space, the orbits of the comets in Oort's Cloud would remain stable for ever. However, the gravitational action of nearby stars periodically perturbs the cloud of comets and, although some are eliminated from the system, many have their perihelion distances reduced so that they enter our solar system within the influence of the planets. A few approach close enough to the Sun to render themselves visible as 'new' comets. Here the word 'new' means first arrivals as distinct from comets which have made several perihelion passages and may be of shorter period. To be consistent with the average number of new long period comets which we actually observe, Oort's Cloud must contain a total of 100 thousand million comets. The amazing fact is that the combined mass of all these comets together is probably little more than that of a moderately large planet like Uranus. Recent work has shown that predicting a theoretical radius for the Oort Cloud based on stellar perturbations alone, gives a value of about 100,000 AU, which is a factor of two too large. However, by including the perturbing effects of the vast, nearby interstellar dust and gas clouds, this radius is reduced to about 50,000 AU, as all comets outside this limit would be completely lost from the Oort Cloud.

Theories about the formation of the Oort Cloud of comets are much more speculative. It has been suggested that the cloud of comets represents an aggregation of interstellar dust collected by the Sun into its own very extended sphere of influence, as it pursues its orbit within the Milky Way galaxy, occasionally passing through one of the vast gas and dust clouds which abound near the galactic plane. This scheme would imply that the planets and comets have completely different origins, despite the great numbers of similarities between the two classes of bodies. Alternatively, perhaps comets were formed in the outer regions of the primitive solar nebula and were subsequently ejected into the Oort Cloud, by a combination of planetary and

Fig 1.4 Distribution in values of 1/a for 86 comets with nearly parabolic orbits.

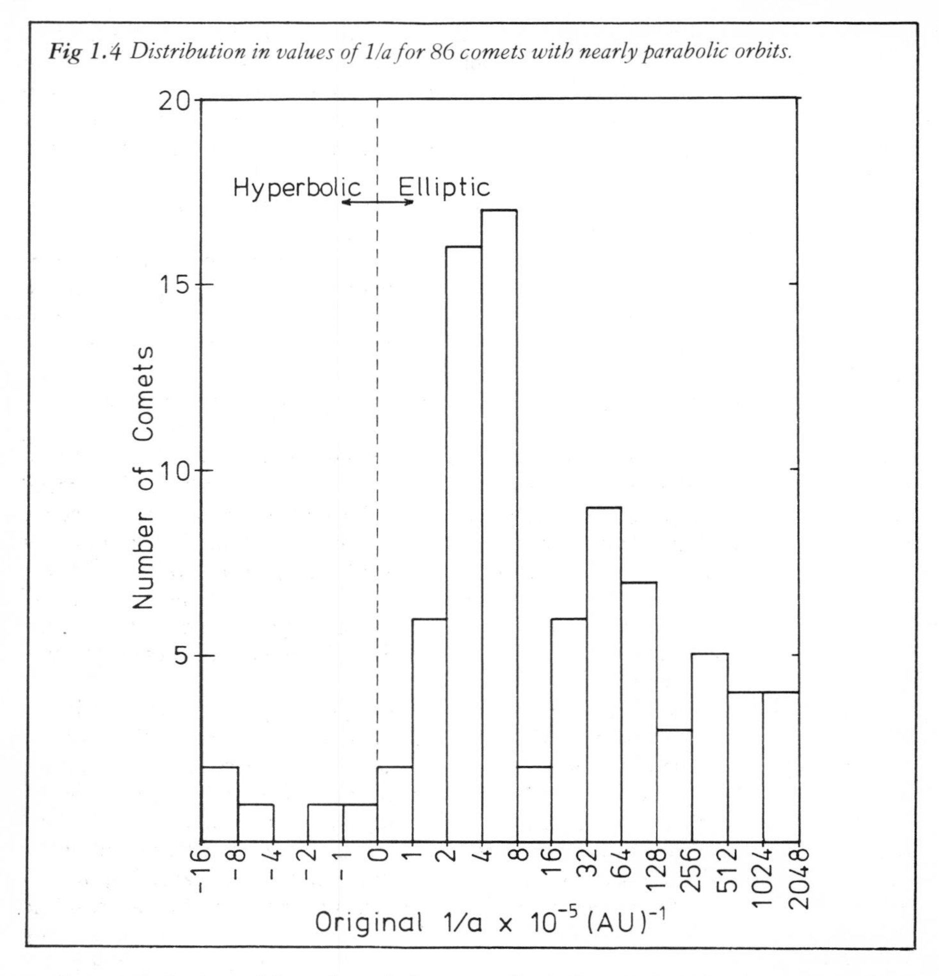

stellar perturbations. Nowadays, it is generally believed that the formation of the cometary cloud in some way was more likely a by-product of the actual formation process of the solar system, and that comets have therefore always been members of that system, and have a common origin with the other bodies within it.

The 'new' and very long period comets move in orbits having inclinations which are nearly randomly distributed, and there is a roughly equal number of comets having direct and retrograde orbits. A retrograde orbit has an angle of inclination i greater than 90°. However, the orientations of the orbits of the long period comets in space are not completely random, as definite groups of objects are noticeable by their similar orbital characteristics. One such group known as the Kreutz Sun-grazing comets is very evident, in that the spatial directions of their perihelia are very similar, and all have perihelion distances of between 0.005 and 0.009 AU, well within the Sun's corona. The property of generally random orbital inclinations for the long period comets having parabolic or nearly parabolic orbits is in marked contrast to the orbits of the short period comets having return periods of less than 30 years. All of these move in direct orbits only, with a mean inclination to the Earth's orbital plane of only 13°.2. The problem now arises as to how the random inclinations of the long period comets become so markedly regular as one passes to the short period ones. This difficulty has been investigated by a number of astronomers and, although the problem is far from solved completely, much progress has been made.

The usual approach is to start with the known cometary orbit and, using a high-speed computer, investigate what happens to this orbit, by calculating either forwards or backwards in time, over as many as a thousand consecutive returns of the comet. In this way, it is possible to assess how the effects of planetary perturbations transform the initial comet's orbit as time passes. It has been found that single, close encounters of a comet to a planet, in particular Jupiter, are inadequate to produce a significant number of short period, low inclination orbits from the initially long period ones, having random inclinations. Far more likely, the capture of a comet into a short period orbit is the cumulative result of many such interactions over a long period of time. Quite obviously, the most massive planet, Jupiter, is a major factor in this mechanism. Furthermore, nearly half of the new comets are so highly perturbed during their very first passage that they leave the solar system for ever on hyperbolic orbits. In general, the number of comets remaining after N perihelion passages within the solar system is proportional to $1/\sqrt{N}$. Dr Edgar Everhart, of the University of Denver, has made much progress in the study of the evolution of cometary orbits, and has accumulated significant evidence that new comets from the Oort Cloud would be able to replenish the number of observable comets, as these were lost through the various perturbations.

So far, we have considered the orbits of comets to precisely follow one or other of the conic sections. Since every body in the solar system attracts every other body in accordance with the law of gravitation, it is evident that no comet can precisely follow one of the conic sections, but it is constantly being disturbed. For example, we may regard the orbit at any instant as an ellipse, but it is a variable ellipse, and different orbital parameters must be specified for different dates. Within the solar system, the main disturbances in the motion of comets are always caused by Jupiter and Saturn, except in the special case where a comet makes a very close approach to one of the other planets. The main effect of such a perturbation is to cause an acceleration in the comet's motion along the line comet to planet. However, this does not generally mean that the comet will move in this direction, because we must combine this small acceleration with the usually far greater attraction between the comet and the Sun. As a result, there may arise various changes in the orbit of a comet. The precise effect of these perturbations will depend not only on the severity of the attraction, governed by the closeness of the approach to the major planet, but also on the time over which the greatest attraction lasts. Hence, the comet which remains in Jupiter's vicinity for some time, as may occur for a short period comet near aphelion, will suffer severe perturbations. The effects are clearly less for a comet whose orbit is highly inclined to the plane containing the planets, since in this case perturbations will only occur in the event of a close approach near one node of the comet's orbit, ie, where the two orbital planes intersect. Similarly, a comet with retrograde motion will suffer smaller perturbations than one moving with direct motion, since the two bodies are passing each other in opposite directions.

One of the best known cases where a cometary orbit was altered very noticeably by Jovian perturbations, is that of Lexell's Comet. In 1770, this comet in its approach towards perihelion, passed between the satellites of Jupiter, and later, also came within $2\frac{1}{2}$ million km of the Earth, on July 1 of the same year. The orbit, which was found to have a period of 5.6 years, was investigated by Lexell, who showed that the comet had been highly perturbed by Jupiter in May 1767, when its orbit had been changed from a much larger ellipse to its present shape, which explained why it had never been seen previously. A second close approach to Jupiter occurred in 1779, and the comet was never seen again, presumably due to resulting large changes in its orbital period and perihelion distance. Lexell's Comet is another example of how the man who calculated the orbit actually got his name associated with the comet; it was in fact discovered by Charles Messier. Not all such encounters with the major planets have such drastic consequences as those portrayed by Lexell's Comet. More common is a rather smaller change in the orientation of the comet's orbit, such as that shown by the comet P/Grigg-Skjellerup. This comet, discovered by Grigg in 1902, was not seen again until 1922, when it was rediscovered by Skjellerup. During this time, the orbit was severely perturbed by Jupiter, and after 1922, it remained stable for the next eight revolutions. In 1964, further perturbations were experienced, producing an orbit with a period of 5.12 years, and a perihelion distance of only 1.001 AU. It was, therefore, possible for close encounters to occur between the comet and the Earth, leading to the occurrence of a new shower of meteors, as a result of the debris deposited along the comet's orbital path.

Even the most well-known comet of them all, Halley's Comet, is not free from the effects of planetary perturbations, despite the fact that it has a moderately inclined, retrograde orbit, with $i = 162°$. Textbooks on astronomy generally give its return period as 76 years, but the precise period may vary by as much as $2\frac{1}{2}$ years either side of the mean value, because of perturbations by the major planets. Indeed, if one plots a graph of the return period of Halley's Comet (between consecutive perihelion passages) against date, over the last 1500 years (see Fig 1.5), these variations are easily seen. The shortest value is only 74.42 years (1835 to 1910) and the largest value is 79.25 years (451 to 530). It will be obvious from the foregoing, that the orbits of nearly all comets are constantly undergoing changes, some rather gradual, others more sudden and extreme. It is therefore rather fortunate that nowadays astronomers possess sophisticated electronic computers with which they can calculate very accurately the expected paths of comets, taking into account the perturbations of all the nine major planets. It is a far cry from the efforts of dedicated workers like Joseph de Lalande and his staff when computing the date of return of Halley's Comet in 1759. For six months the team calculated from morning to night, and sometimes even at meals. They calculated the distance of each of the two planets, Jupiter and Saturn from the comet, separately, for every degree, for 150 years. They predicted that Halley's Comet would reach perihelion on April 13 1759. The actual date of perihelion passage was March 13 1759, only 31 days earlier, no less a tribute to the skill of the mathematicians.

The tremendous accuracy attainable by the astronomical mathematicians, even at the beginning of the 19th century, was demonstrated by the surprising discovery that comets appear to defy the Newtonian Law of Gravitation, the very theory which had previously been triumphantly proven by earlier calculations of cometary orbits. In 1819, J.F. Encke of Germany was studying the motion of a short period comet, discovered by 1786, and which possessed the shortest period (3.3 years) of any comet known. To his surprise, Encke found that the comet persisted in returning at each revolution about $2\frac{1}{2}$ hours too soon. This may seem like a very trivial

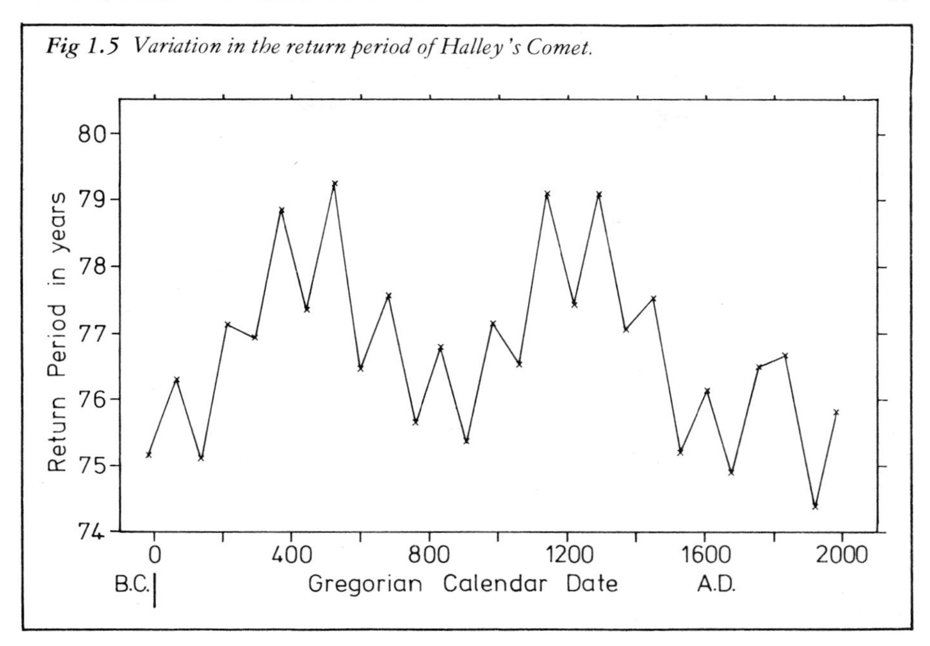

Fig 1.5 *Variation in the return period of Halley's Comet.*

deviation, but the accuracy of the observations of the comet's positions placed this discrepancy well beyond the possible errors of measurement and calculation. Comet Encke, as it is now known, still continues to return sooner than predicted by pure Newtonian theory, to this day. Even Halley's Comet has similar strange tendencies, only it has persisted in arriving an average of 4.1 days late during its past 11 apparitions. In fact the majority of the short period comets show deviations from pure Newtonian motion, and this phenomenon is now known to be caused by a non-gravitational force, which can speed up or slow down the motion of the comet in its orbit. To understand this fascinating anomaly, it is important to consider the exact nature of the cometary nucleus and its chemical composition.

CHAPTER 2

THE ANATOMY OF COMETS

The space-age astronomer observes comets by many techniques, apart from visual observations with the naked eye, binoculars or telescopes. This is because we now know that the visual wavelengths of comet spectra (ie, those which can be detected with the human eye) and on which the bulk of our knowledge rests, are only a very small part, not necessarily representative, of the total spectral information actually available. Photographs may be taken not only at visual wavelengths, but also in the higher energy, shorter wavelength ultra-violet wavebands. It is possible to measure the brightness of comets (a process known as photometry) at both visual wavelength and in the lower energy, longer wavelength infra-red regions of the spectrum. By analysing the spectrum of the light received from the comet, astronomers can discover the actual chemical composition of comets by a technique known as spectroscopy. Furthermore, since 1969, scientists have made observations of comets from space, above the thick layers of the Earth's atmosphere. This includes ultra-violet observations from rockets and orbiting spacecraft, as well as observations by astronauts orbiting the Earth. A particular example of this was the important series of photographs of Comet Kohoutek 1973 XII, by the American astronauts, Carr, Gibson and Pogue, in the Skylab Space-Station, one of which is shown in Fig 2.1.

Our current knowledge of the structure of comets points to the existence of a small central body known as the nucleus, from which all cometary material, both gas and dust, originates. Unfortunately, no photograph of a nucleus has ever been secured, but it is probably irregular in shape, with a radius ranging from just a few hundred metres to perhaps a few tens of kilometres. The mass of a nucleus is estimated as between 10^{11} kg and 10^{16} kg, so even the most massive is only one hundred-millionth the mass of the Earth. The average density would be no greater than 2 grams per cubic centimetre, and perhaps rather less, depending on the exact composition. At a reasonable distance from the Sun, say beyond 5 AU, the nucleus, composed of a roughly equal mixture of water-ice and dust, merely absorbs energy from the sunlight falling upon it, and this energy received, simply warms the surface of the nucleus. At this stage, the nucleus is normally invisible to us and it is inactive.

As the cometary nucleus approaches the Sun, the temperature of the surface layers increases sufficiently for the water-ice contained within the nucleus to sublimate. The process of sublimation is where a substance is converted directly from a solid (in this case ice) to a vapour, by direct heating. For water-ice the temperature of the sublimating layer would be about $-60°$ Centigrade. Most of the energy received by the nucleus goes into the sublimation of ices. If the main constituent of the nucleus is

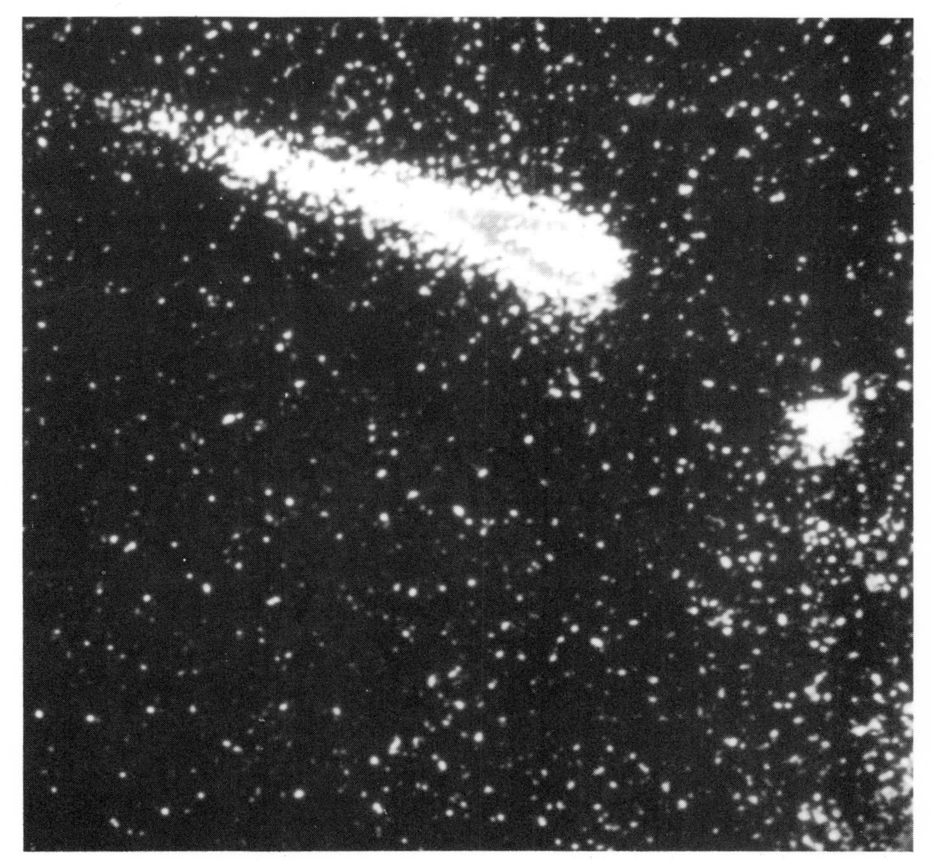

Fig 2.1 *Comet Kohoutek from Skylab.*

indeed water-ice, then the onset of cometary activity should occur at a distance of about 3 AU from the Sun. This activity is shown by the formation of an essentially spherical cloud of gas and dust surrounding the nucleus, known as the coma. The coma is formed by substances like water, ammonia and methane, streaming away from the nucleus at typical speeds of a few hundred metres per second. The gas flowing away from the nucleus picks up dust particles through frictional forces and pulls them away from the nucleus. The observation that gaseous neutral molecules, other than H_2O (from the water-ice) are present when the coma first appears, supports the theory that the water-ice contained in the comet's nucleus is a special type known as a clathrate hydrate. This rather grandiose name means that there are cavities present in the crystalline ice in between the water molecules bound in the crystal. These cavities may contain completely different chemical substances as minor constituents. When the water-ice sublimates, these minor constituents will be released at the same time. In some cases, if there is an excess of these minor constituents within the water-ice, the sublimation process may no longer be controlled by the water. This means that sublimation of the minor constituents could occur well beyond the distance of 3 AU (quoted above), as the comet approaches the Sun. This

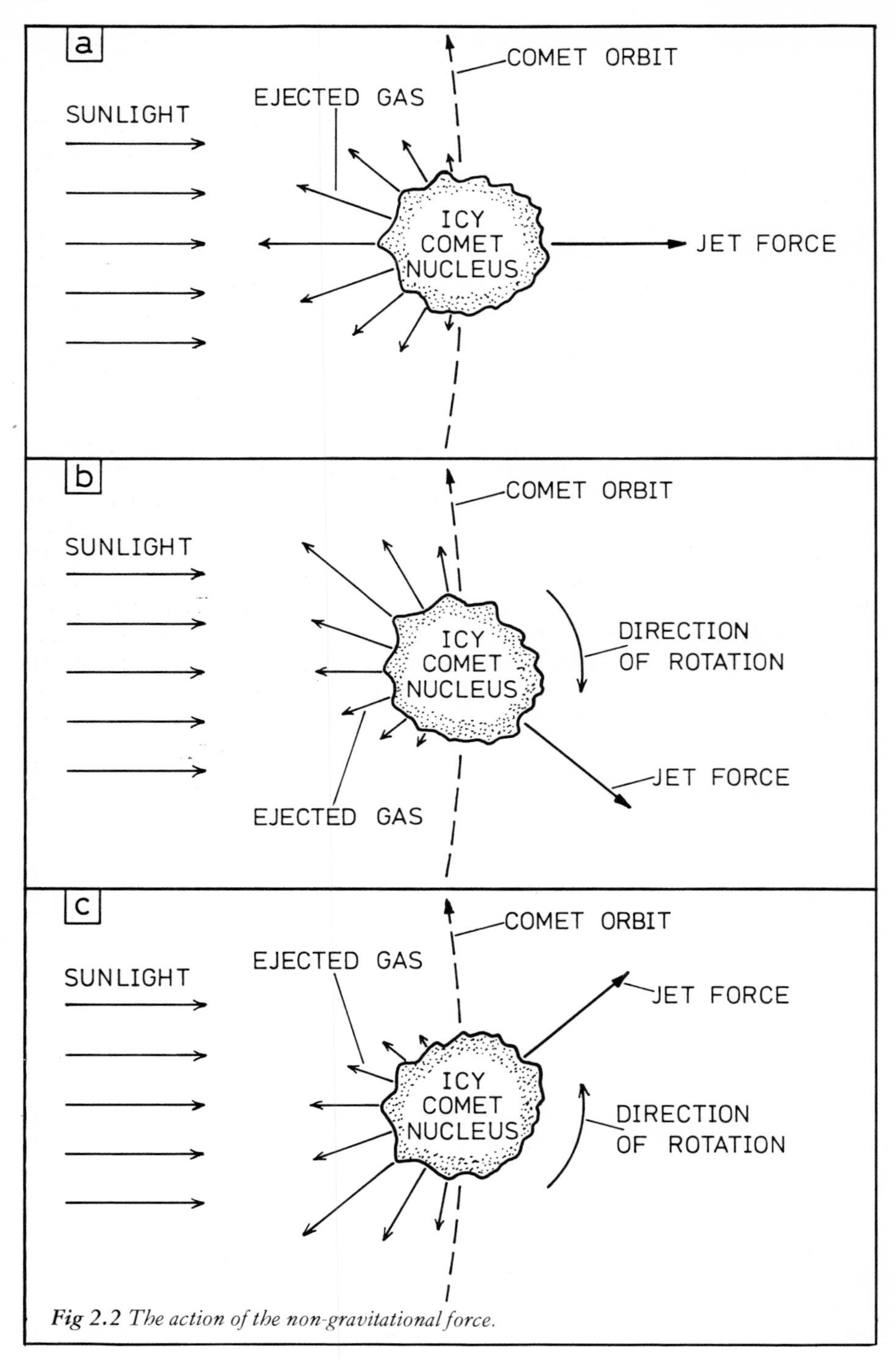

Fig 2.2 *The action of the non-gravitational force.*

explains why some comets appear much brighter than expected at greater distances from the Sun. Comet Kohoutek, for example, was first seen while still 4.9 AU from the Sun.

There is one very important result of the process described above, whereby ices sublimate from the cometary nucleus when subjected to sufficiently intense solar radiation, and drag away some of the entrapped dust particles. The result is a reactive force on the nucleus, directed away from the Sun, and it is this which causes the non-gravitational motion of comets first discovered by Encke in about 1820, and described in the previous chapter. The thrust of escaping gases on the side of the nucleus nearest to the Sun can alter the orbit and hence the period of the comet. The three diagrams shown in Fig 2.2 illustrate the action of the non-gravitational force. If the nucleus is not rotating (see Fig 2.2a) the sublimation of surface ices will generate a small but steady jet force that pushes the comet radially outwards from the Sun. In this case, the orbital period would not change from one revolution to the next, although the comet subjected to such a force would follow a slightly larger orbit than a comet in which the force was absent. Unfortunately, this difference in the size of the orbit is below the limit of detectability. However, a rather different picture emerges if the cometary nucleus is rotating. In fact, rotation appears to be a universal property of all astronomical bodies and, hence, it is to be expected for cometary nuclei. If the nucleus rotates in a direction opposite to its direction of motion around the Sun (see Fig 2.2b), there will be a time lag in the sublimation rate, as the sunlit hemisphere turns around. In these circumstances, the resulting jet force possesses a component which will oppose the comet's motion, thereby slightly reducing its orbit and causing the comet to arrive early at its perihelion passage. Conversely, if the

Fig 2.3 *The change in the period of Comet Encke.*

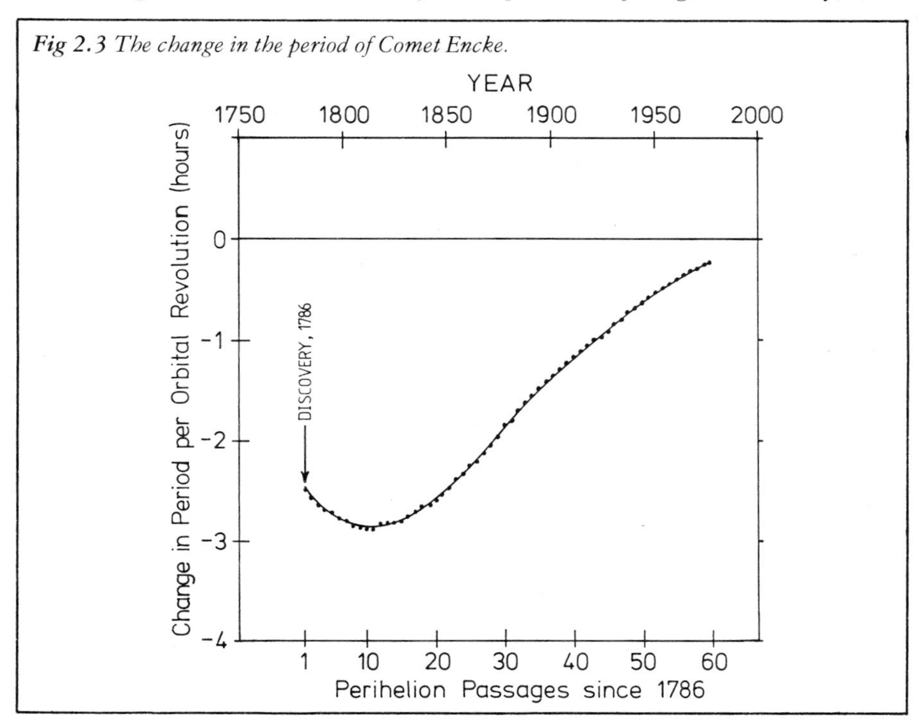

comet is rotating in the same direction as that in which it moves around the Sun (see Fig 2.2c), the component of the jet force will accelerate the comet's motion, thereby expanding its orbit and delaying its arrival at perihelion.

Some very comprehensive studies of the non-gravitational forces in comets have been made. It has been shown that in addition to the direction of rotation of the nucleus, the orientation of the spin axis of the tiny nucleus in space is also of great importance. An analysis of 37 short period comets indicated no significant systematic orientation of the axis of rotation with respect to the orbital plane. However, a detailed study of the short period comet P/Encke revealed remarkable changes in the non-gravitational force acting on that comet's nucleus. From 1819 to 1825, the drag force gradually increased but then, after reaching a peak, declined in a fairly uniform fashion by a factor of more than eight times. At the end of this period, instead of the comet consistently returning $2\frac{1}{2}$ hours earlier every revolution, today, the decrease in orbital period amounts to only a few minutes per revolution. This variation is shown in Fig 2.3. A simple explanation is that the resisting jet force has simply decreased over this period. If this is so, then why did it rise to a maximum in 1825, and then fall? It is now thought that the real reason lies in the changing direction of the spin axis of the nucleus. This would affect the geometry of the jet force and so explain the observations. If the nucleus of Comet Encke is not a perfect sphere, but an oblate spheroid, the body will rotate about its shortest axis. This geometry indicates that the jet force will tend to tip the axis of rotation in one direction or another, and the oblate comet nucleus will precess rather like a spinning top or gyroscope. Calculations by Sekanina and Whipple conclusively proved that between 1786 and 1977, in 59 perihelion passages, the axis of rotation of the nucleus had gyrated more than 100° across the sky. This gyration completely accounts for the observed changes in the orbital period.

Another interesting aspect of comet nuclei is the method used to determine their rotational periods, again devised by Whipple. Many comet nuclei develop active regions on their surface which eject gas when the rotation presents this particular area towards the Sun. The result of these periodic ejections of gas is a series of concentric or nearly concentric haloes formed in the diffuse coma enveloping the nucleus. The haloes are formed at intervals, which coincide with the rotation period of the nucleus. It is here that the drawings of comets, made before the invention of photography, can assist in the study. Comet Donati, the beautiful comet visible in 1858, and depicted in Fig 2.4, is the best example of this. From drawings made on successive days, measurements of the spacing of the haloes showed that the comet was rotating with a period of 4.6 hours. The haloes were generated repeatedly for a period of three weeks. A similar, but less accurate, analysis of drawings of Comet Encke, gave a probable value of $6\frac{1}{2}$ hours for the rotation period of its nucleus.

Apart from the cometary haloes, observers have meticulously recorded many other phenomena in the coma envelopes of comets, including fan-shaped rays, jets and streamers. These asymmetries in the coma of a comet, with respect to the nucleus, particularly in the inner regions, demonstrate that the sublimation process does not occur uniformly over the surface of the nucleus. Many comet nuclei do possess small active regions which dominate the sublimation process when they are turned towards the Sun. Such observations have been of great value in determining both the orientation of the spin axis and the rotation period of the cometary nucleus.

Even the nucleus of Halley's Comet is rotating. We know that because the comet consistently returns to perihelion an average of 4.1 days late every apparition, its

Fig 2.4 *Donati's Comet.*

nucleus must be rotating in the same direction as its motion around the Sun. That is to say, the thrust of the escaping gases will reach a peak in the 'afternoon' of the comet's 'day'. The resultant jet force pushes the comet forwards in its orbit, the nucleus drifts outwards from the Sun, its orbital period increases and it returns later than predicted. Analysis of the non-gravitational parameters affecting the orbit of Halley's Comet between 1607 and 1911 have shown that the spin axis of the nucleus is stable. It shows no obvious precessional motion and the spin axis is not lowering or lifting up with respect to the orbital plane. Furthermore, it appears that the comet's ability to outgas has not changed substantially over that period. Following a careful study of observations of Halley's Comet made at previous apparitions, F.L. Whipple has been able to show that the rotation period of its nucleus is 10 hours 19 minutes.

As a comet approaches the Sun, a greater number of neutral gas molecules flow away from the nucleus, dragging many dust particles with them and increasing the size of the coma. The nucleus and coma combined, are called the head of the comet. The essentially spherical coma may extend from 10,000 km to as far as 1 million km from the nucleus. The largest coma ever recorded was probably that of the Great Comet of 1811, which shone very conspicuously during the autumn of that year and remained visible for many weeks. Its coma had a diameter of about 2 million km, considerably larger than the Sun.

Some comets are predominantly dusty and, as a result, most of the light we receive from them is simply sunlight scattered by the myriads of tiny dust particles. In this case, the spectrum of the head of the comet is a continuous spectrum of reflected sunlight. Other comets contain little dust and, since the molecules and atoms in a gas scatter light only feebly, such gaseous comets become bright only through the action

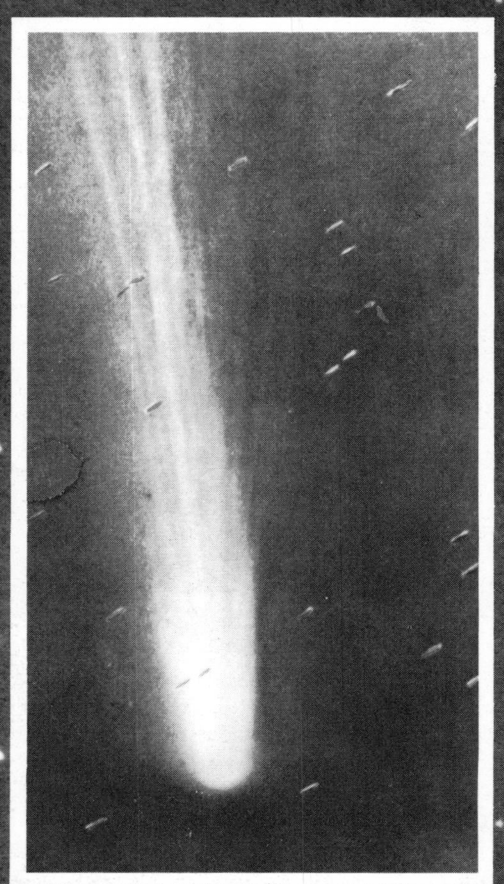

Left *Fig 2.6 Comet Mrkós of 1957 from Byurakan, USSR.*

Left *Fig 2.7 Comet Arend-Roland, April 24 1957* (E.M. Lindsay).

Fig 2.5 *West's Comet, March 9 1976, (04.45–05.00 UT; 05.00–05.15 UT composite)* (photograph by Paul Doherty).

of other processes. Firstly, highly energetic ultra-violet radiation emitted by the Sun pulls the gas molecules apart and dissociates them into similar constituents. For example, water (H_2O) would be dissociated into atomic hydrogen (H) and the hydroxyl radical (OH). The atoms, or broken molecules, are then excited to fluoresce by solar radiation, that is, they absorb the solar light at one wavelength and then re-radiate it either at the same wavelength or, more usually, at a series of longer wave-lengths having less energy. A major part of the light received from gaseous comets, emanates from such bands of wavelengths. They are mostly emitted by the dissociated molecules of the light elements—hydrogen (H), carbon (C), nitrogen (N) and oxygen (O). Emission bands have been observed from radicals such as CH, NH, NH_2, CN, C_2 and C_3. Observations of comets using millimetre wavelength radio emissions have revealed the existence of the stable molecules hydrogen cyanide (HCN) and methyl cyanide (CH_3CN) within the coma. These two substances were first noted in Comet Kohoutek in 1974. In addition to the molecular emission bands, the yellow lines of sodium metal often appear if a comet approaches to within 0.7 AU of the Sun, and occasionally much closer still, many spectral lines of the heavier elements, potassium, calcium, iron, nickel and chromium are seen. With the exception of these atomic lines, the elements H, C, N and O are the major contributors to the cometary spectrum. The fact that the heavier atoms are only observed when comets are very near the Sun, implies their release from some of the larger dust particles, and this has been attributed to vapourisation of the dust.

The existence of atomic hydrogen in comets was an important discovery. If ordinary water-ice is the major component of the comet's icy nucleus, then bright comets should be surrounded by huge clouds of hydrogen (H) and the hydroxyl radical (OH), arising from the dissociation of water molecules (H_2O). A hydrogen atom in space emits its fluorescent radiation not in the visible region of the spectrum but at far ultra-violet wavelengths, at a particular line in the spectrum named Lyman-alpha. This particular emission was first detected from space, in 1969, in Comet Tago-Sato-Kosaka 1969 IX. This comet was found to be surrounded by a hydrogen cloud approximately 1.6 million km in diameter. The following year, a similar cloud was found around Comet Bennett, and since that time, several comets have been shown to possess giant hydrogen comae, extending to as much as 10 million km in diameter. Radiation was also detected from the hydroxyl radical, in much larger quantities than was first thought. There seem to be roughly equal numbers of hydrogen atoms and hydroxyl radicals, confirming the theory that water-ice is a major constituent of comet nuclei.

Comet West, discovered in August 1975, and a prominent naked eye object during March 1976, also possessed a large hydrogen coma. It was interesting, in that it was an unusually dusty comet, and it will be mentioned later in connection with the disruption of cometary nuclei. Comet West was a beautiful object, and is depicted in Fig 2.5. When a bright comet such as this reaches a distance of only 1 AU from the Sun, it emits hydrogen at a rate of anything between 500 kilograms per second (kg/sec) to 1,300 kg/sec. These figures should be compared with the corresponding total rate at which mass is being lost by the comet, of between 9,000 kg/sec and 24,000 kg/sec (24 tonnes/sec).

As a comet moves in towards perihelion, the dust particles are carried away from the nucleus in all directions by the continual outward flow of gases. Their velocities and directions of emission are combined with the motion of the comet, to produce their final orbital paths. The dust particles liberated from the nucleus are of two

***Fig 2.8** Comet Humason; 48-inch Schmidt 1959* (Hale Observatories).

types. The smaller particles, having diameters of a few tenths of a micron (one micron equals one millionth of a metre) to several microns, and mean densities of the order of one gram per cubic centimetre, are blown in the anti-solar direction by solar radiation pressure. They move away from the nucleus on hyperbolic orbits. With respect to the comet, the orbit of each dust particle curves backwards from the anti-Sun line and direction of motion of the comet, because of the conservation of angular momentum. The result is a curved dust tail, otherwise known as a Type II or Type III tail, which is often fairly broad and rather diffuse, ranging from one million to ten million km in length. The particles in the dust tail are probably silicate in composition, and they are visible only through the reflecting and scattering of sunlight. The large dust particles expelled from the region of the nucleus, having diameters from one micron up to a few tens of microns, are relatively unaffected by the force of solar radiation pressure, and they go into their own orbits around the Sun. It is these particles that scatter sunlight, to produce the zodiacal light. This material is also dispersed along the comet's orbit to produce a meteor shower, if ever it enters the Earth's atmosphere.

The ejection velocity of any dust particle from the comet's nucleus depends on both the size and density of the particle (ie, its mass), and the diameter of the parent comet's nucleus. For a particle density of about 0.4 grams per cubic centimetre, and a comet nucleus 6 km across, the ejection velocity ranges from 100 m/sec for a particle of 0.1 micron diameter, down to only 35 m/sec for a particle of 10 microns in diameter, and even less for larger particles. Because ejected particles have to escape from the comet's gravitational pull, there is an upper limit on the maximum size of particle which can actually escape from the nucleus. As the maximum emission of gas from the nucleus occurs near the comet's perihelion passage, this is also the place where the greatest quantity of dust is introduced into the comet's vicinity. For

Comet Bennett, seen in 1970, the rate of dust production was calculated as 20 tonnes/sec at a distance of 0.56 AU, assuming a nuclear diameter of 5.2 km. A fine example of a dust tail was shown by Comet Mrkos 1957 V, depicted in Fig 2.6.

Occasionally, dust tails are seen which appear to point in the sunward direction, in apparent contradiction to the general property of dust tails pointing away from the Sun. These so-called anti-tails are not truly material ejected towards the Sun, but appear as a result of a projection effect of sunlight reflecting dust grains distributed along the orbital plane of the comet. An anti-tail is only visible when the Earth is near the plane of the comet's orbit and the conditions are exactly right for sunlight to catch the dust grains at the correct angle. As the Earth approaches this plane, the anti-tail first becomes visible as a stubby tail pointed at a considerable angle to the comet-Sun direction. At the moment the observer crosses the orbit plane, the anti-tail becomes extremely thin and narrow, resembling a spike and attaining its greatest length. Thereafter, the anti-tail continues to swing around becoming initially rather short and stubby and later, broad and fan-shaped. It is possible to determine the time of orbit crossing quite precisely, by measuring the position angle of the anti-tail with respect to the comet-Sun direction. One of the most spectacular anti-tails ever seen was exhibited by Comet Arend-Roland 1957 III, whose anti-tail attained a greatest length of 14° on April 24 1957, and this is shown in Fig 2.7. This comet was extremely dusty, and calculations have shown that the dust grains in the anti-tail were emitted from the comet between 60 and 80 days before the Earth crossing. It appears that the dust grains present in anti-tails are rather larger than those in the normal dust tail, having diameters of several tens of microns and, rarely, up to a millimetre in size.

Most comets, particularly the brighter ones, display a second type of tail, quite different from the dust tail. These tails are almost straight and, generally, much narrower and larger than the dust tails, their lengths ranging from 10 million to 100 million km. They are called plasma tails, otherwise known as Type I tails. Both kinds of tail may exist separately, or together in the same comet. Both point, generally, away from the Sun. Like the gas within the coma, plasma tails emit light by fluorescence, not by scattering sunlight. The gas molecules responsible for this radiation differ from those in the coma in that they are ionised. In other words, electrons have been removed to leave the molecules with a net positive charge. These ionised molecules are not generally formed until a comet is within 1.5 AU of the Sun. The dominant visible ion is ionised carbon monoxide (CO^+), the emission of which is responsible for the blue colouration of the plasma tails. Several other ionised molecules are generally present in the plasma tail, namely, CO_2^+, N_2^+, H_2O^+, and the radicals OH^+ and CH^+. No un-ionised molecules or radicals are found. Although plasma tails are not usually observed in comets beyond about 1.5 AU from the Sun, an exception was Comet Humason 1959 X, which possessed a plasma tail but virtually no dust at its very great perihelion distance of 4.3 AU. This rather faint comet is shown in Fig 2.8.

The extremely tenuous plasma tails are formed as a result of a fairly complex interaction between the comet and something termed the solar wind. The Sun continuously ejects millions of tonnes of gas per second. This solar wind flows away from the Sun at an average velocity of 400 km per second, and has a temperature of about 1,000,000°C. The solar wind carries with it a magnetic field generated by currents of electrons flowing in the gas. It is these high energy electrons which actually ionise the molecules in the coma (along with the solar radiation) in the first place.

Fig 2.9 Comet Morehouse 1908.

The ionised cometary molecules are accelerated by the force of the solar wind to velocities often exceeding 100 km/sec, and they are swept by it in a direction away from the Sun. The ionised molecules are also trapped and concentrated along the magnetic field lines carried by the solar wind, and this results in the material in plasma tails being concentrated into thin bundles or streamers. Additional highly complicated structures may be present in the plasma tail, in the form of knots and kinks, which appear to move along the tail away from the nucleus. These may persist for a few hours and, very rarely, for a day. The ever-changing structures seen in plasma tails are undoubtedly produced by short-term irregularities in both the solar wind and the comet's activity. Shock waves caused by solar flares may produce pronounced kinks in the plasma tail of a comet, or even a spiral helix structure. In rare cases it is possible for the polarity of the magnetic field to change suddenly, and this results in detachment of the plasma tail from the comet, while a new tail is forming. This particular phenomenon occurred with the rather beautiful Comet Morehouse 1908 III, shown in Fig 2.9.

The tremendous variety of cometary phenomena is obvious from what has been stated so far. Even with our considerable knowledge of their properties, comets are still notorious for their erratic and unpredictable behaviour. In particular they are sometimes subject to unexpected variations of brightness. Several comets have exhibited outbursts resulting in an increase in their brightness of a hundred times or more. The most persistent example of this is Comet P/Schwassmann-Wachmann 1, which orbits near Jupiter, with a period of about 16 years, and undergoes outbursts at reasonably frequent intervals. In late May and early June 1973, Comet P/Tuttle-Giacobini-Kresák flared by some ten thousand times in brightness, only to fade back into obscurity within three weeks. All comets show this type of behaviour to some degree.

Sometimes comets are observed to split into several components, accompanied by an outburst in brightness. The velocities of separation of the components are probably no more than a few tens of metres per second. Only for the Kreutz Sun-grazing family is the cause clearly evident. These comets approach the Sun so closely at perihelion, that it is possible for the Sun's gravitational pull to disrupt the nucleus into several components by tidal forces. The other notable split comet was P/Biela, discovered by Montaigne, in 1772. It behaved normally until early 1846, when the comet was observed to elongate and then separate into two distinct parts, which slowly moved apart. At the next return in 1852, their separation had increased to nearly 2 million km. No sign of either component has been seen since, except for meteor showers irregularly dispersed near their orbit and producing fine displays in 1872 and 1885. More recently, the brilliant Comet West, already mentioned in this chapter, was seen to break up into no less than five separate fragments, between March 8 and March 24 1976.

The material used to form the coma, hydrogen cloud, dust and plasma tails, is forever lost to a comet. It is estimated that about 1 per cent of a comet's mass may be lost during each revolution, mostly near perihelion. This means that for a small comet having a diameter of 2 km, a layer several metres thick is removed during each passage. Hence, when the sublimation process has continued for many orbits, as in the case of short period comets, the supply of ices will run out and a 'dead' comet will result. However, estimations of cometary life-times are still open to considerable error. One would expect a comet to grow steadily fainter as it grew older but, in fact, there are very few comets for which secular changes in intrinsic brightness can be confidently stated. In particular, Encke's Comet has been observed over 59 revolutions since 1786, and the estimated mass loss is much less than 1 part in a 1,000 per period, and we may expect it to survive for many more than 100 orbits.

So far, comets have only been considered rather generally. In introducing the main subject of this book, Halley's Comet, it is worth remembering that all the characteristics considered; their motion, physical properties, chemical composition, and eccentric behaviour, apply as much to Halley's Comet as they do to any other. Halley's is the only really active comet which has a well-determined orbit and has shown reasonably reliable behaviour for at least 2,000 years. With a period of approximately 76 years, it has apparently made relatively few returns to the solar neighbourhood since being perturbed into our solar system from the Oort Cloud. It may, therefore, be one of the newest comets with a period less than 200 years, and it is certainly the most active short period comet known, exhibiting a wide range of cometary phenomena, having a large dense coma, both dust and plasma tails, and possessing jets and streamers. All other short period comets have a more limited range of activity, and appear to be near the end of their physical evolution. Although other short period comets have highly evolved nuclei and inner comae, they cannot be expected to have highly active nuclei or well developed tails. Studies of Halley's Comet therefore enable us to compare the properties of a bright and active comet at successive appearances.

CHAPTER 3

EDMOND HALLEY: ASTRONOMER ROYAL

Edmond Halley, the second Astronomer Royal, was born in 1656, at Haggerston in Outer London. His life extended through one of the most dramatic periods in the history of science, and he made a great personal contribution to it. Yet, ironically, he is always remembered as lying in the shadow of his ever greater contemporary, Isaac Newton. But for Halley's connection with the famous comet which bears his name, he would be barely remembered at all except by historians of astronomy. Fate can sometimes be very unfair.

He came of a good Derbyshire family. His grandfather, Humphrey, had been a prosperous haberdasher, and his father, also named Edmond, carried on a successful business in London. The future astronomer was one of three children, but little is known about the other two. The girl, Katherine, seems to have died in infancy; his brother Humphrey certainly died in 1684, but otherwise our ignorance is complete. We are not even sure of the date of Edmond's birth. In one of his surviving notes he gave it as October 29 1656. The difficulty about this is that his parents had been married only seven weeks earlier. There are two charitable explanations for this discrepancy. The date may have been wrongly given; alternatively, the actual marriage may have taken place well before the church ceremony.

From an early age young Edmond showed that he was capable of great things. There was no shortage of money, and by 1671 he was captain of St Paul's, one of the best schools of the day. He began his scientific work while still in his teens, and in 1673 he entered Queen's College, Oxford. A brilliant career seemed to lie ahead of him.

He was, moreover, one of those people who is almost universally liked. He had a jovial disposition, and was an amusing companion; he was entirely free of malice or jealousy, and in this he was very different from some of his contemporaries, notably John Flamsteed and Robert Hooke. His marriage, in 1682, was apparently a happy one; his wife, née Mary Tooke, died six years before him, and he also survived his only son, Humphrey, who died in 1741.

There are many anecdotes about him. For instance, he certainly knew Czar Peter of Russia when that somewhat ferocious ruler came to England in 1698, ostensibly to learn about shipbuilding. It is said that after a far from teetotal evening, the Czar climbed into a wheelbarrow, and in the subsequent ride Halley pushed him through a holly hedge. In fact it is unlikely that this story is true, and more probably it comes into the Canute-and-the-waves category, but Halley would have been quite capable of such an escapade. In 1703 John Flamsteed made the rather jaundiced

***Fig 3.1** Edmond Halley.*

remark that 'Halley now talks, swears and drinks brandy like a sea-captain'.

Yet of Halley's ability and scientific dedication there was never any doubt, and it was his good fortune to live through a period of change. When he was born, there were still some eminent scientists who believed the Sun to be in orbit round the Earth rather than vice versa. By the time he died, in 1742, the old idea of a central Earth had been abandoned even by the diehards, mainly because of Isaac Newton's work upon the laws of universal gravitation—and the immortal *Principia* was published at Halley's urging and with Halley's money.

The great revolution in thought had been sparked off in 1543 by Mikołaj Kopernik, always remembered by his Latinised name of Copernicus, who was as unlike Halley as could possibly be imagined. He was a Polish cleric with an astronomical bent, and he realised that the old theory of a central Earth surrounded by a revolving sky simply did not fit the facts. He therefore dethroned the Earth from its proud central position, and put the Sun there instead. To be frank, this was his only major contribution, and most of the rest of his theory was wrong, but he had taken the vital step. Prudently, he refrained from publishing it until the end of his life, because he foresaw trouble from the Church authorities—and in this he was certainly wise. As long afterwards as 1600 one of his followers, Giordano Bruno, was burned at the stake in Rome for daring to teach the Copernican theory. (This was not Bruno's only crime in the eyes of the Church, but it was certainly a major one.) There was also the case of the great Italian, Galileo Galilei, who was the first astronomer to make serious use of a telescope. In 1610 he made a series of remarkable discoveries, all of which convinced him that the Sun, not the Earth, was the centre of the planetary system. He saw the four bright satellites of Jupiter, the phases of Venus, spots on the Sun and the almost innumerable stars making up the Milky Way; he was also the true founder of the science of experimental mechanics, and he was not afraid to speak his mind. Eventually the Church acted. Galileo was summoned to Rome, tried by the Inquisition, and forced into a public but completely hollow recantation. This, it is worth noting, happened only two decades before Edmond Halley was born.

Meanwhile, there had been theoretical developments. The best observer of pre-telescopic times was Tycho Brahe, a brilliant if eccentric Dane who built an observatory on the Baltic island of Hven and drew up a catalogue of stars which was by far the best of its time. When he died, in 1601, his observations came into the possession of his last assistant, Johannes Kepler, who used them well—but not in the way that Tycho would have hoped. Tycho himself could never believe that the Earth was anything but the most important body in the universe, and he also supported the old maxim that all orbits must be circular, because the circle is the 'perfect' form, and nothing short of perfection could be allowed in the heavens. Kepler was more flexible in his outlook. Using Tycho's careful measurements of the movement of the planet Mars, he established that the planets do indeed move round the Sun, but in ellipses rather than circles. Kepler's three Laws of Planetary Motion, the last of which was published in 1618, formed the basis of all future work.

Yet though Kepler had explained how the planets move, he had no clear idea of why they do so; this was left to Newton, almost 70 years later. Meanwhile telescopes were being steadily improved, and telescopic sights were introduced, so that it became possible to draw up really accurate star catalogues. This was of more than academic importance, since observations of the Moon and stars provided the only practicable method of marine navigation.

Britain has always been a seafaring nation, and depends largely upon its ships. Unfortunately, finding one's position when out of sight of land presented serious problems. Two things have to be known: latitude and longitude. Latitude is relatively easy to measure to an observer north of the equator, because all that has to be done is to measure the altitude of the Pole Star and then make a minor adjustment to allow for the fact that the Pole Star is not exactly at the true celestial pole (at present it is rather less than one degree away). The altitude of the pole will always be equal to the observer's latitude. Thus if the altitude is found to be 51°, the observer's latitude will be 51° north. It was longitude-finding which gave the real trouble, and on more than one occasion British ships were lost because they turned east instead of west, or vice versa.

Something had to be done, and in Britain a Board of Longitude was set up to tackle the problem. The obvious solution was to measure the local time (which was easy enough) and then compare it with Greenwich time; the difference between the two would depend upon the longitude of the observer. But at that time there was no clock which could keep time accurately enough when carried on board ship. The only alternative was to use the Moon as a sort of clock-hand. If its position could be measured against the background of stars, Greenwich time could be worked out. This involved using a really precise star-catalogue, and even Tycho's was not good enough.

The Royal Society had been founded in 1660, due largely to the encouragement of that much-maligned monarch Charles II, and various schemes were considered. The method of 'lunar distances' seemed to be the only hope, and accordingly the King ordered that a new observatory should be set up in order that a new star catalogue could be compiled. In 1675 the Royal Greenwich Observatory was completed, with buildings designed by Sir Christopher Wren, who had been professor of astronomy at Oxford before turning to architecture. The Rev John Flamsteed was put in charge, subsequently assuming the title of Astronomer Royal, and work began. In the end Flamsteed succeeded in his aim, though not without many delays and tribulations in which both Newton and Halley became involved.

Greenwich lies more than 50° north of the equator. Therefore, stars near the south celestial pole can never be seen, because they do not rise above the horizon, and a large area of the sky is inaccessible. This was where the youthful Halley decided to take a hand. At the age of 20, even before he had taken his degree, he left Oxford and set out for St Helena, 16° south of the equator. His decision was no boyish whim; he planned to chart the southern stars, and he had the support both of his father (who provided enough money to buy the necessary equipment) and his university tutors. By 1677 he was installed upon that somewhat bleak and inaccessible island. The climate was unfavourable, and there were also brushes with the local governor, whose name was Gregory Field and who seems to have been a most objectionable person; but Halley did all that he had hoped, and by the time he left St Helena, over a year later, he had drawn up a good catalogue of over 300 stars.

Fig 3.2 Frontispiece of Principia.

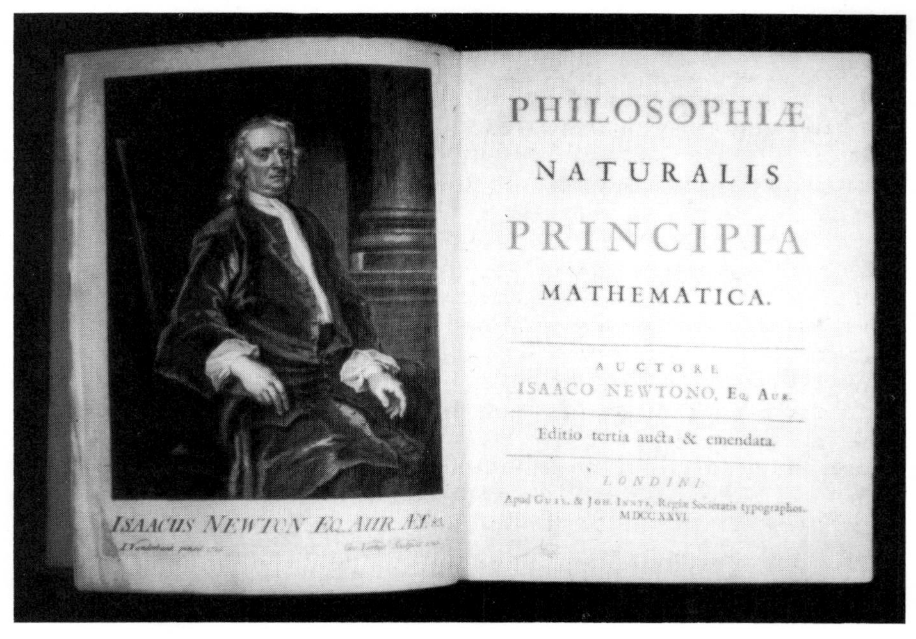

His reputation was made; he was awarded an honorary degree from Oxford, and he was also elected a Fellow of the Royal Society. He was even nicknamed 'the southern Tycho'.

By then he was in regular correspondence with Flamsteed, Newton, Hooke, Wren and other leading scientists; he made a trip to the city of Danzig (now Gdansk) to see the famous astronomer Hevelius, and his friendly personality made him welcome everywhere. He was an observer as well as a theorist, and he was particularly interested in two bright comets which appeared soon after his return from the south, one of which was seen in 1681 and the other in 1682. But the most important development of this period came in the year 1684, which in other respects was a sad one for him; his father died in mysterious circumstances (possibly murdered) and his brother Humphrey also died. It was just as well that he had plenty of scientific work to occupy him.

At the rooms of the Royal Society, Hooke, Wren and Halley discussed the problem of gravitation, and in particular what we now term the inverse square law. Hooke had published some papers about it, and was certainly on the right track. It can be explained by simple arithmetic, so let us take the convenient case of two imaginary planets which move round the Sun at distances of 2 million km and 5 million km respectively. (No planet is anything like as close-in as this, but let us make things as easy as we can.) Two squared, or 2×2, is 4. Five squared is 25. Then according to the inverse square law, the Sun's pull upon the two planets will be in the ratio of $\frac{1}{4}$ to $\frac{1}{25}$, so that the force on the distant planet will be only $\frac{4}{25}$ of that on the closer. In this case, it can be shown that each planet will move in an elliptical orbit.

The law seemed to be valid, but there was no firm mathematical proof, and the only man capable of working it out was Newton. Halley knew this quite well, and so he went down to Cambridge to talk matters over. He was astonished to find that Newton had solved the problem some years earlier, but had not published it, and had even lost his notes!

Halley used all his powers of persuasion, and eventually Newton agreed to re-work the calculations. During 1685 and 1686 he spent all his time in writing the book which he called the *Philosophiae Naturalis Principia Mathematica* (Mathematical Principles of Natural Philosophy), but which is always remembered simply as the *Principia*. It has been described as the greatest mental effort ever made by one man, and it ushered in what may be termed the modern phase of astronomy. Halley was determined to see it published, and even paid for it out of his own pocket. All in all, it is probably true to say that this was his greatest contribution to science.

It may be asked why the Royal Society did not finance the book. The answer was simple: there was no money available. All the reserve funds had been spent upon producing a huge book by Francis Willughby called *The History of Fishes*. No doubt this book was a learned tome, but as a publishing venture it was disastrous, and there were many copies left unsold. In fact Halley, who had become a salaried official of the Society (thereby temporarily forfeiting his membership) was later presented with 50 copies of the book in lieu of arrears of salary. Just what he did with them is not on record.

By now Halley's reputation was second to none. He was amazingly versatile; his papers ranged over many fields—for instance, *A Discourse Tending to Prove at what Time and Place Julius Caesar made his first Descent upon Britain; A Way of Estimating the Necessary Swiftness of the Wings of Birds to Sustain their Weight in the Air*; and *A Method of Walking Under Water* (the latter being a description of a primitive diving bell; in 1691 Halley made some practical experiments with it at Pagham, on the Sussex coast not far from Selsey Bill). Another of his interests was archaeology, as witness his paper *The Ancient State of the City of Palmyra* in the Middle East. However, astronomy was always his first love, and he produced an important paper in 1691 in which he described a method of measuring the length of the astronomical unit. Today the distance is known very accurately; it is given as 149,597,870 km or 92,957,209 miles. In Halley's day the uncertainty was considerable.

Kepler's Laws had made it possible to work out the relative distances of the planets from the Sun compared with that of the Earth. In fact, it was easy enough to draw up a scale model of the solar system, and the problem lay in finding out one absolute distance—it did not matter which. Halley proposed to make observations of the transits of Mercury or Venus across the face of the Sun. Such transits do occur occasionally (only with these two planets, of course; all the others move beyond the Earth's orbit), and Halley found that by measuring the exact times of these transits, as seen from different points on the surface of the Earth, the actual distances could be found. Mercury transits more often than Venus, but it is less suitable, because it is smaller and further away. Transits of Venus occur in pairs, each pair being separated by over a century; thus there were transits in 1631, 1639, 1761, 1769, 1874 and 1882, while the next will be in 2004 and 2012. Halley had no hope of living long enough to see one, but his method was put into practice even though it did not prove as satisfactory as had been hoped. As the whole method is now completely obsolete there is no point in describing it further, and the next transits will be regarded with no more than passing interest, but it is worth recalling that Captain Cook's famous voyage of exploration in 1769 was undertaken primarily to carry official astronomers to the South Seas in order to observe the transit.

Halley himself made three sea voyages, though for a different reason; his aim was to chart magnetic variation—that is to say, the difference between true north and magnetic north. In 1698 he was put in command of the pink *Paramore* (sometimes spelled *Paramour*) and went to South America and the West Indies. His First Lieutenant apparently objected to being put under the orders of a landlubber, and eventually Halley took over the navigation personally, bringing the *Paramore* home with no trouble at all. The second voyage, beginning in 1699, took him to the far south—to the region of the Falkland Isles and South Georgia. The third, in 1701, was restricted to the Channel area. His studies of magnetism proved to be invaluable, and for this alone he deserves to be remembered. After his last voyage there was no further use for the faithful *Paramore*, which was eventually sold in 1706 for the princely sum of £122. During this period Halley resigned his post of Clerk to the Royal Society, and was duly re-elected to full Fellowship. In 1704 he also became Savilian Professor of Astronomy at Oxford.

Meanwhile, the storm-clouds were gathering over Greenwich.

John Flamsteed was an excellent observer, but by nature he was touchy and stubborn, and he also suffered from ill-health, so that he and Halley could not be expected to be kindred spirits. Moreover, Flamsteed was a perfectionist. He disliked publishing any of his measurements until he was absolutely confident about them, and when Newton and others asked for his results he took their requests as personal affronts. This time there was no financial problem; Prince George of Denmark, husband of the new sovereign Queen Anne, had volunteered to pay the entire cost of publication, but Flamsteed remained obdurate, and there was a great deal of ill-feeling.

Finally, Flamsteed gave the Royal Society committee a copy of his observations as well as an incomplete manuscript of the catalogue itself. He stressed that the observations could go forward, but not the catalogue. Printing was duly begun. Four years later Flamsteed had still not produced the finished catalogue, and by then the committee had lost patience. In 1711 the storm broke, with the publication not only of Flamsteed's observations, which he had authorised, but also of the catalogue, which he had not. Halley had not only supplied information on his own account, but had also written a preface to which Flamsteed took the strongest possible exception. In 1715 he managed to secure a number of copies of the book, and publicly burned them 'that none might remain to show the ingratitude' of two of his countrymen—by which he meant Newton and Halley. He never did see the final catalogue; he died in 1719, and the work was finished and edited by two of his last assistants, Crosthwait and Sharp.

Halley has been sharply criticised for his part in the affair, and it has been said that he went behind Flamsteed's back. Actually this is incorrect. Halley kept Flamsteed fully informed about what actions were to be taken, and there was no hint of underground manoeuvring, but it was all too clear that relations between the two great astronomers had been permanently severed. It was perhaps ironical that on Flamsteed's death, Halley was at once appointed Astronomer Royal in his place.

By now he was well over 60, but his energy was as great as ever. Unfortunately he had problems to face. King Charles' generosity in building the Observatory (financed, typically, by the sale of 'old and decayed gunpowder' to the French) had not extended

Fig 3.3 *Old Greenwich.*

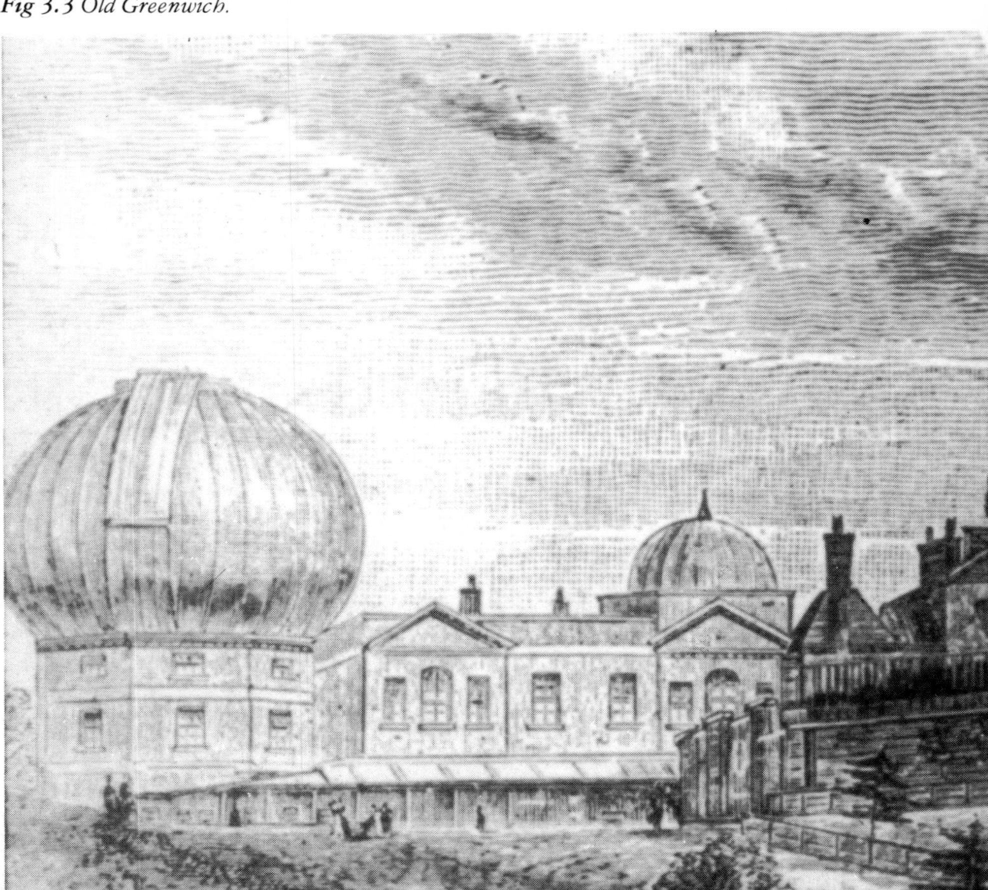

to providing instruments, and these had been paid for by Flamsteed himself, so that technically they were his property. When Halley arrived, he asked Mrs Flamsteed to remove herself and her belongings as soon as possible. This was done with devastating promptness. The good lady took not only her personal effects, but also the telescopes and measuring instruments, so that Halley was left with the shell of an observatory and nothing more. It was a tragedy for science; the original instruments have never been traced, and it is hardly likely that they will ever be found now. Halley had to begin again.

As soon as the new equipment had arrived, he began a series of observations of the movements of the Moon. This was essential for navigational purposes, but Halley knew that it would take him 18 years, and it is greatly to his credit that he carried the programme through to the end. It is true that Flamsteed's catalogue and Halley's lunar positions were never fully used for their original purpose—longitude-finding—because of the invention of the marine chronometer, which kept time so well that the lunar distances method became obsolete even before it could be put into practice, but the measurements were none the less valuable for that. And among Halley's other achievements must be mentioned that of his discovery of stellar proper motions; he established that three bright stars, Sirius, Procyon and Arcturus, had shifted appreciably since ancient times, so that the old term of 'fixed stars' was inappropriate.

Halley remained active and healthy until very near the end of his life in 1742. Only during the last two years did he show any sign of failing physically, and his mind remained clear. Typically, his last action, only minutes before he died, was to call for a glass of wine—and drink it.

CHAPTER 4

'ONE AND THE SAME . . .'

Edmond Halley's first acquaintance with a bright comet came in 1680, when a particularly imposing visitor appeared in the sky. It was discovered by Kirch at Coburg, in Saxony, on November 14 of that year, and passed through perihelion on December 18. It had already developed a considerable tail, up to 30° long, and when it reappeared in the evening sky around December 20 it was magnificent. The tail extended over a vast arc, attaining a length of 90°, and curved so that it was likened to an enormous sabre. The nucleus seemed to change from night to night; telescopically it was compared with 'a burning coal', and there are several reports indicating that the colour was indeed pronounced, though admittedly the redness may have been somewhat exaggerated. During the first weeks of January 1681 it dominated the night sky, and it faded slowly; it was last seen on March 19 by Newton, using a telescope of 7 ft focal length. Halley followed it closely, and was suitably impressed. How exactly was it moving?

Fig 4.1 *Great Comet of 1680.*

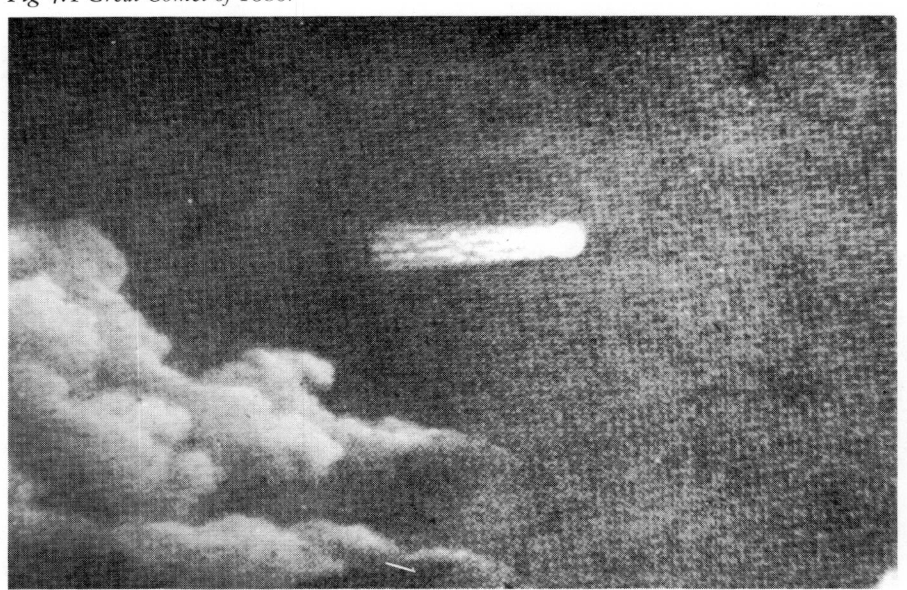

Ancient astronomers had believed comets to be local; the Chaldaeans, who were among the first to divide up the stars into groups or constellations, believed them to be fires produced by eddies in violently-rotating air, while Aristotle, the great philosopher who died in 322 BC, was convinced that they were atmospheric phenomena. Ideas of this kind were killed off by Tycho Brahe, who made a careful study of the bright comet of 1577 and concluded, quite rightly, that it must be much further away than the Moon. But whether or not comets were true members of the solar system remained questionable. Kepler tended to believe that they travelled in straight lines. Alternatively, it was suggested that they might follow open curves (parabolae). The problem was not easy to solve; at that time a comet could be followed over only a short arc of its orbit, under which circumstances it was difficult to tell a parabola from an ellipse or even a circle.

At first Halley was open-minded. When he tried to calculate the orbit of the comet of 1680 he found that its path was much the same as those of comets seen previously in 43 BC and 531 AD and 1106. Could it be that these were different apparitions of one and the same comet, moving round the Sun in a period of around 575 years? Admittedly he realised that the evidence was slender, because the earlier comets had not been accurately plotted; but his compatriot, the Rev William Whiston, was much less cautious, and worked backwards to establish that the comet had approached the Earth closely in early times and had been the direct cause of Noah's Flood! Actually, we now know that the true period of the comet amounts to thousands of years, and we can have no real idea of when it will return; but meanwhile the comet of 1682 had appeared, and proved to be much more informative.

It was discovered in August, by a German astronomer named Dörffel, and was seen shortly afterwards both at Paris Observatory and from Greenwich. Halley was not at Greenwich at that time, but Flamsteed was, and found that by the 21st of the month the tail had stretched out to a full 10°. (Remember that the apparent diameter of the full Moon is only about half a degree.) On September 9 Flamsteed recorded that the head was 'dull', and barely visible in the twilight sky. Halley saw the comet on the following night, but by then the whole head had become blurred.

From the mainland of Europe the comet made a better showing. At the end of August, Picard, at Paris, found that the head was about equal to a star of the second magnitude (that is to say, as bright as the Pole Star), and from Danzig Hevelius wrote that the comet could then be seen all night, with a tail over 15° in length. Hevelius also described a luminous 'ray' thrown out from the head into the tail, and that there were times when the tail itself did not point directly away from the Sun. There were various other observations of the comet, which was presumably less brilliant than that of 1680, but was nevertheless a prominent naked-eye object for some weeks. The date of the perihelion passage was August 24.

All this, of course, took place before the publication of Newton's *Principia*, and it was some time before Halley started to make exact calculations. When he did, he began on the assumption that the orbit must be parabolic, but before long he was struck by something very significant indeed. The orbit was remarkably similar to those of comets previously observed in 1607 and in 1531. In each case the inclination to the ecliptic was between 17° and 18°; the perihelion distance from the Sun was between 0.5 and 0.6 AU, and the direction of motion was retrograde—that is to say, opposite in sense to that of the Earth. Checking further back in time, Halley came across comets seen in 1456, 1378 and 1301, again with similar elements. He was in no hurry to make any hard and fast predictions, but by 1705 he felt reasonably confident, and he published a book, *Astronomicae Cometiae Synopsis* (Astronomical Synopsis of Comets) in which he forecast that the comet would return once more in or about 1758.

He knew, of course, that a comet is of low mass, at least by planetary standards, so that it would be very subject to perturbations by giant planets such as Jupiter and Saturn. Therefore, the period would not be constant, and 76 years would be only an average. He wrote:

'Now many things lead me to believe that the comet of the year 1531, observed by Apian, is the same as that which in the year 1607 was described by Kepler and Longomontanus, and which I saw and observed myself at its return in 1682. All the elements agree, except that there is an inequality in the times of revolution; but this is not so great that it cannot be attributed to physical causes. For example, the motion of Saturn is so disturbed by the other planets, and especially by Jupiter, that its periodic time is uncertain to the extent of several days. How much more liable to such perturbations is a comet which recedes to a distance nearly four times greater than that to Saturn, and a slight increase in whose velocity could change its orbit from an ellipse into a parabola? The identity of these comets is confirmed by the fact that in 1456 a comet was seen, which passed in a retrograde direction between the Earth and the Sun, in nearly the same manner; and although it was not observed astronomically, yet from its period and its path I infer that it was the same comet as that of the years 1531, 1607 and 1682. I may, therefore, with some confidence predict its return in the year 1758. If this prediction is fulfilled, there is no reason to doubt that other comets will return.'

He was suitably modest. 'You see, therefore, an agreement of all the elements in these three, which would be next to a miracle if they were three different comets; or, if it was not the approach of the same comet towards the Sun and Earth in three different revolutions, in an ellipsis around them. Wherefore, if accordingly to what we have already said, it should return again in the year 1758, candid posterity will not refuse to acknowledge that this was first discovered by an Englishman.'*

One might have imagined that an announcement of this kind would cause great excitement in astronomical circles. In fact it did not—at least, not at the time—because 1758 was a long way in the future. Halley himself had no hope of living to see whether or not his prediction would be fulfilled; he would by then have reached the advanced age of 102. But as the years passed by, interest grew, and the centre of activity shifted from England to France.

One of the leading mathematicians of the time was Alexis Claude Clairaut, who had been something of a child prodigy, and was busy studying mathematics when he was 11, publishing his first paper at the age of 13. In 1757 he decided that he would compute the effects of the known planets upon the wandering comet, and do his best to find out just when and where it would reappear. He did not work alone; he was joined by Joseph Jérôme de Lalande, who already had a great reputation as an astronomer, and by Madame Jean André Lepaute, wife of a clockmaker. Of her, Clairaut commented that 'although she was not pretty, she did have an elegant figure, and a pretty little foot'.

The French mathematicians had left things rather late, because the comet was almost due, and they had to work quickly. 'During six months,' wrote Lalande, 'we

* Halley's *Tabulae Astronometricae*, published in 1749, seven years after his death. This quote is taken from the 1752 edition, which had an English translation added.

calculated from morning to night, sometimes even at meals; the consequence of which was, that I contracted an illness which changed my constitution for the rest of my life. The assistance rendered by Madame Lepaute was such that without her we should never have dared to undertake this enormous labour; in which it was necessary to calculate the distance of each of the two planets, Jupiter and Saturn, from the comet, separately for every successive degree, for 150 years.' They finally found that the comet would be delayed for 100 days by Saturn and for no less than 518 days by Jupiter. This would mean that perihelion would occur on April 13 1759.

They were only just in time. When they were ready to announce their results it was already November 1758, and the hunt was on. One of the most eager of all the hunters was another Frenchman, Charles Messier. The latter was no mathematician; he was purely an observer. At the age of 14, in 1744, he observed a brilliant comet, and thenceforth comets dominated his life. He was employed by Nicholas de l'Isle, Astronomer of the Navy in Paris, and carried out his work with small telescopes, mainly from the turret of the Hôtel Cluny—hardly a suitable site, but better than nothing at all. De l'Isle had made some calculations of his own, and had decided that the comet would probably become visible about 35 days before perihelion. Unfortunately he was not sure when perihelion would occur, and Messier began to search in the indicated area as early as 1757, using a small reflector of focal length 4½ ft.

For many months he worked away without result, keeping rigorously to the regions of the sky which de l'Isle had indicated. By bad luck the Paris skies were generally cloudy towards the end of 1758, and Messier had to make the most of the few clear nights, but on January 21 1759 he made the great discovery. In his own words:

'The whole days was very fine and without cloud; in the evening I went over the sky with the telescope, keeping to the limits of the two ovals drawn upon the celestial chart which was my guide. By about six o'clock I discovered a faint glow resembling that of the comet I had observed in the previous year; it was the Comet itself, appearing 52 days before perihelion! There is cause to presume that if M de l'Isle had not made the limits of the two ovals so restricted, I would have discovered the comet much earlier, while it had a greater elongation from the Sun. It had been much closer to the Earth two months before 21 January.'

It is easy to picture Messier's feelings—but his elation was short-lived. De l'Isle flatly forbade him to announce his discovery; Messier was to continue observing, but that was all. 'I was a loyal servant of M de l'Isle,' he commented wryly. 'I lived with him at his house, and I obeyed his command.' It must have been a blow; but all he could do was to follow the comet until it vanished into the twilight as it neared the Sun.

Even worse, Messier did not have priority after all. On Christmas Night 1758 a Saxon amateur astronomer, Johann Palitzsch, had picked up the comet using a telescope of much the same size as Messier's. This time there was no delay; Palitzsch made his announcement at once, and before long other German observers had provided full confirmation, though the news took some time to percolate through to France. Messier and de l'Isle heard about it at the end of March, by which time the comet had reappeared and had been seen by both Messier, on the 31st, and La Nux, five days earlier.

At last, on April 1, de l'Isle made the news of Messier's independent discovery known. Not surprisingly, the astronomical world was sceptical. What had been the point of keeping silent? Even today it is hard to give an answer, and one must agree with the English astronomer J.R. Hind, who wrote that 'such a discreditable and selfish concealment of an interesting discovery is not likely again to sully the annals of Astronomy'. Certainly it was something which Messier could never forget, even though it was true that de l'Isle had given him his first astronomical post. There was another ironical twist of fate ahead. Messier continued to hunt for comets almost to the end of his life in 1817, and discovered more than a dozen; but he also drew up a catalogue of star-clusters and nebulae, not because he was interested in them, but because they could only too easily be mistaken for comets. He made his list simply to avoid wasting time on them. Today it is by this catalogue that Messier is remembered—not for his comets!

Palitzsch, incidentally, was a farmer who lived at Prohlis near Dresden. He was a skilled amateur astronomer who was known to have exceptionally keen sight, and he had been conducting a systematic search. The first confirmation came from a professional, Dr Hoffman, on December 28.

The date of perihelion proved to be March 13 1759, so that the French mathematicians had been in error by about a month, which was a remarkably good result. Obviously they had no calculating machines to help them; such things lay in the far future. Neither did they know of the existence of the two outer giant planets, Uranus and Neptune, which were not discovered until 1781 and 1846 respectively.

It must be admitted that the 1759 return was not spectacular. Before perihelion the comet was faint, and there is no record that it was seen with the naked eye before perihelion passage. (The oft-quoted story that Palitzsch discovered it without optical aid is certainly untrue.) After it reappeared it was somewhat brighter, and during April and May it was well seen from Europe, though the best views of it were obtained from the southern hemisphere. The maximum length of the tail was given as 47°. Clearly it was not so brilliant as it had been in 1682, or as it was to be again at the return of 1835.

Generally speaking, a comet is known by the name of its discoverer or discoverers, but there are occasions when the name chosen is that of the mathematician who calculates the orbit. Halley's Comet comes into this category. So does Encke's Comet, first seen in 1786 by Messier's friendly rival Pierre Méchain, which was found to be periodical much later by Johann Encke of Berlin and has now been seen at over 50 returns. But Encke's Comet is puny indeed compared with Halley's, and in fact Halley's Comet is the only bright member of the class whose period is short enough for us to be able to predict it.

By now many periodical comets are known, with periods ranging from 3.3 years (Encke) to 164.3 years (Grigg-Mellish, last observed in 1907). But the 1759 return of Halley's Comet marked a major step forward in our understanding of what we now call celestial mechanics, and it ensured that the name and achievements of the second Astronomer Royal will never be forgotten.

CHAPTER 5

HALLEY'S COMET IN HISTORY

Comets can be spectacular objects. True, there has been a comparative dearth of brilliant visitors during our own century, and there have been no comets as magnificent as, for instance, those of 1811, 1843 and 1861. But when comets really do put on good displays they cannot pass unnoticed, and it was natural for early star-gazers to record them.

Greek astronomy did not begin before the time of Thales of Miletus, who lived around 600 BC. Before that we depend almost entirely upon observations made by the Egyptians and, particularly, the Chinese. Eclipses were regarded as very important (the Chinese attributed them to the activities of a hungry dragon which was trying to gobble up the Sun), but comets too were carefully recorded. Inevitably, some of these ancient records refer to Halley's Comet.

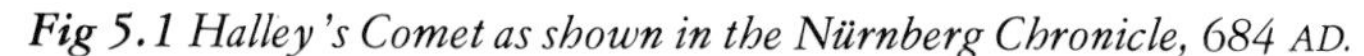

***Fig 5.1** Halley's Comet as shown in the Nürnberg Chronicle, 684 AD.*

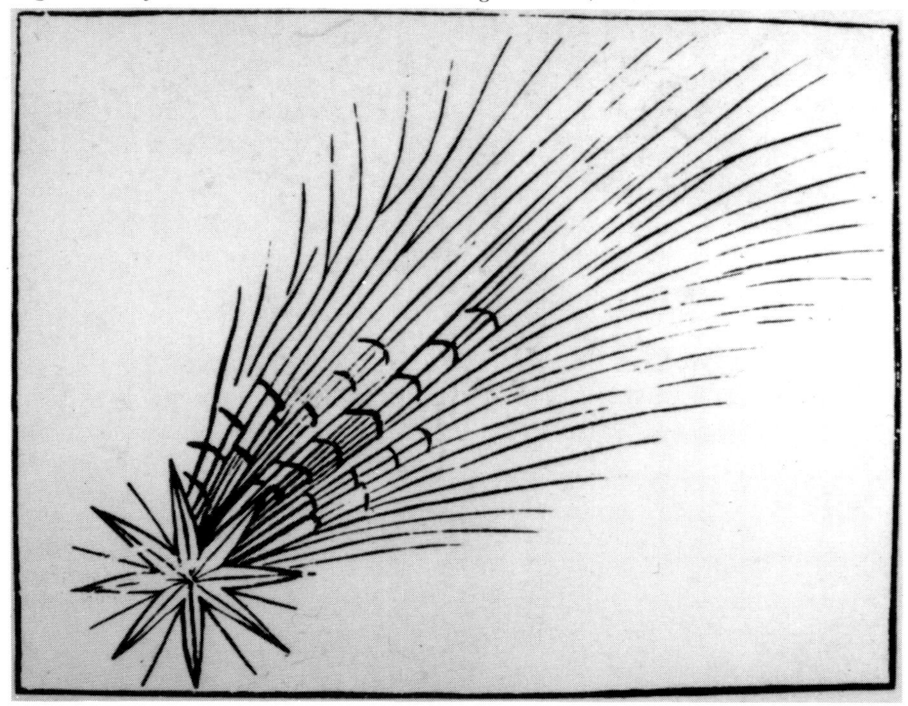

It is far from easy to make exact identifications, because a comet—even a bright one—is of low mass, and is only too easily perturbed by the actions of the planets. Edmond Halley was able to make his great discovery because the movements of the comets of 1531 and 1607 had been reasonably well observed, but the returns of 1759 and 1835 made is possible to improve the accuracy of the calculations, and by 1852 a well-known English astronomer, John Russell Hind, had traced the story of Halley's Comet as far back as 12 BC.* Further work was done just before the 1910 return by Cowell and Crommelin, at Greenwich, and by now we can follow the records back to the return of 240 BC. There is a chance that a comet seen in 467 BC from China was also Halley's; M.H. Vilyev of the USSR has even attempted to go back to 622 BC, and there is a possibility of an observation in 1059 BC, but we cannot be sure. Moreover, there are differences of opinion regarding the dates of the earliest returns; we have given them here according to the official list by Yeomans and Kiang, published in April 1981 by the Royal Astronomical Society, and the years as 164, 87 and 12 BC, but other earlier texts give 163, 86 and 11 BC, a discrepancy of one year, caused by the changing of the calendar. Here then is the list of returns since 240 BC:

Year	First seen	Last seen	Perihelion
240 BC	—	—	May 25
164 BC	Not recorded	Not recorded	Nov 12
87 BC	Aug	Aug	Aug 6
12 BC	Aug 26	Oct 20	Oct 10
66	Jan 31	Apr 11	Jan 25
141	Mar 26	May	Mar 22
218	Apr	May	May 17
295	—	May	Apr 20
374	Mar 3	May	Feb 16
451	June 10	Aug 16	June 28
530	Aug 28	Sept 27	Sept 27
607	Apr 18	July	Mar 15
684	Sept 6	Oct 24	Oct 2
760	May 16	July	May 20
837	Mar 22	Apr 28	Feb 28
912	July 19	July 28	July 18
989	Aug 11	Sept 11	Sept 5
1066	Apr 1	June 7	Mar 20
1145	Apr 26	July 9	Apr 18
1222	Sept 3	Oct 23	Sept 28
1301	Sept 15	Oct 31	Oct 25
1378	Sept 26	Nov 10	Nov 10
1456	May 26	July 8	June 9
1531	Aug 1	Sept 8	Aug 26
1607	Sept 21	Oct 26	Oct 27
1682	Aug 24	Sept 22	Sept 15
1759	Dec 25 1758	June 22	Mar 13
1835	Aug 5	May ± 19 1836	Nov 16
1910	Aug 25 1909	June 16 1911	Apr 20

* Note that some tables give a year's discrepancy for the early returns. It was only in 1582 that most countries changed to the modern Gregorian calendar, while Britain did not make the change until 1752.

Only the return of 164 BC is not definitely recorded, and even here there are some vague Chinese references to a bright comet seen about that time. So let us now look back and see how Halley's Comet has behaved over the centuries.

467 BC A bright comet was seen from both Greece and China. There is no record of its track across the sky, but the comet may have been Halley's. It is also worth noting that according to Aristotle, a large meteorite fell at Aegospotami while the comet was on view. This may or may not be significant; on the whole it is probably coincidence, since there is no definitely known association between comets and meteorites, but again one can never be sure.

240 BC Chinese records describe a bright comet which was certainly Halley's, but the observations are not precise. All we can really say is that the comet was spectacular enough to cause general interest.

164 BC The only gap in the record.

12 BC (sometimes given as 11 BC). Here we come to the first return about which we have useful information. The consuls in Rome were M. Valerius Messala and P. Sulpicius Quirinus, during whose term of office, according to the historian Dion Cassius, a comet 'was suspended over the city'. From China the comet was carefully followed as it passed from Gemini through Leo, Boötes, Hercules, Serpens and Scorpius; its movement was relatively rapid, because at that time the comet was comparatively close to the Earth. Perihelion occurred in October, and there is every reason to suppose that the comet was bright. It was lost in the Sun's rays two months after it had first been seen.

The reason why this return is so interesting is because efforts have been made to identify Halley's Comet with the Star of Bethlehem. And though there seems no chance now that this is correct, it is worth discussing in slightly more detail. We are handicapped from the outset because we know so little about the Star in the East. It is mentioned only once in the Bible: in the Gospel according to St Matthew, Chapter 2. Many people will know the text: The Wise Men came to Herod saying, 'Where is he that is born King of the Jews? for we have seen his star in the east, and have come to worship him'. Verses 7 to 10 run as follows: 'Then Herod, when he had privily called the Wise Men, inquired of them diligently what time the star appeared. And he sent them to Bethlehem, and said, Go and search diligently for the young child; and when ye have found him, bring me word again, that I may come and worship him also.

'When they had heard the King, they departed; and, lo, the star, which they saw in the east, went before them, till it came and stood over where the young child was. When they saw the star, they rejoiced with exceeding great joy.'

That is all. St Matthew says no more, and the other Gospels do not mention the episode at all, so that our information is hopelessly scanty. One thing we do know for certain is that Christ was not born on December 25 1 AD. Our AD dates are taken from the calculations of a Roman monk, Dionysius Exiguus, who died in the year 556. He gave the date of Christ's birth as 754 years after the founding of Rome, which is, to put it mildly, dubious. Moreover, December 25 was not celebrated as Christmas Day until the fourth century—by which time the real date had been forgotten, so that our Christmas is wrong too.

Almost all authorities agree that Christ was born before 1 AD. The favoured date is 4 BC. This is well after the return of Halley's Comet, but in any case there is a simple, powerful argument to show that the whole idea is completely out of court. St Matthew infers that the Star was not seen by anyone apart from the Wise Men, in which case it cannot have been an ordinary astronomical body. For the same reason we can also dismiss other theories involving novae, supernovae or perhaps a close conjunction of two planets. Even Venus has been put forward as a candidate (and this is brought up again every time Venus is brilliant around the end of December). If the Wise Men were deceived by Venus, they can hardly have been very wise, and Herod would have had to do no more than go and look. Halley's Comet could not have provided the answer even without an unacceptably large discrepancy in date.

66 AD (all our dates from now on are AD). Two comets were recorded by the Chinese about this time, the second of which was probably Halley's. It was also mentioned by Flavius Josephus, a Jewish historian, who referred to a broadsword-shaped star over Jerusalem during the time when it was under siege, and he also recorded a comet 'that stayed a whole year'—an obvious exaggeration.

141 Chinese observations tell us that there was a bright comet in this year, presumably Halley's. It was followed for over a month, and was presumably at least fairly conspicuous.

218 Again we have Chinese records, and also a note by Dion Cassius that the comet was 'a very fearful star, extending its tail from the west eastwards', and travelling through Auriga, Gemini, Leo and Virgo. The death of the obscure Roman emperor Macrinus, which took place at the same time, was naturally blamed on the comet!

295 Chinese observations mentioned the comet, moving from Andromeda as far as Virgo. Little is said about its brightness.

374 Again we depend on the Chinese, who gave a position in the constellation Ophiuchus. It was recorded only after perihelion.

451 Here we are on firmer ground. The comet was recorded not only in China, but also over Europe, and seems to have been prominent as it moved from near the Pleiades through Leo and into Virgo. This was the time of one of the world's decisive battles. The armies of Attila the Hun had been plundering Europe, forcing Theodosius II to pay tribute and penetrating as far as Orleans, where they were met by a force commanded by the Roman general Aetius. The resulting Battle of Chalôns was a disaster for Attila—which on the whole was certainly a good thing, since otherwise it is very likely that the Huns would have overrun the whole of the European mainland.

530 We know very little about this return; Chinese reports are vague, and so are those from Europe. But as the year fits, and there were no other bright comets seen during this period, the identification with Halley seems definite.

607 This is the only AD return about which there is any real doubt, because once more the Chinese provide our only source of information. They observed two bright comets in 607-8, and we cannot be sure which was Halley's, but one of them must have been.

684 Ma-tuan-lin, the Chinese historian, refers to a comet seen in the western sky during September and October. It was also recorded in Europe, and the first known drawing of it refers to this return. The drawing was published in the *Nürnberg Chronicle*—more properly known as the *Liber Chronicorum* or *Weltchronik*, by Hartmann Schedel. It was printed in 1493 and shows woodcuts by the German artist Michel Wohlgemuth and his stepson Mihelm Pleydeburff. A rough representation of the comet, with accompanying text, appears on the page dealing with the year 684—though it is also repeated elsewhere in the book because of the limited printing techniques of the time. The text adds that at this time there was rain, thunder and

lightning, causing the deaths of people as well as animals, while grain withered in the fields and there was an outbreak of plague.

760 The Chinese observed the comet for two months, describing it as being very brilliant and 'like a great beam'.

837 Four comets were seen around this time, but it was Halley's which caused really widespread interest. In fact, this seems to have been the most brilliant return in history. The maximum magnitude of the head has been calculated as -3.5, brighter than any planet apart from Venus, while the tail extended over an angular distance of 93°. The closest approach to Earth occurred on April 11, when the distance was only 0.04 AU; that is to say, about 6,000,000 km. Modern astronomers would have liked the same conditions to apply to the return of 1986!

912 Little positive information; the comet was seen only during July, and was much less brilliant than it had been in 837.

989 Not well recorded, though the Chinese saw the comet and it was also noted by several Saxon writers, such as the historian Elmacin.

1066 A famous return, because of its association with the Battle of Hastings. It also provides us with our first contemporary picture of the comet.

The apparition was a favourable one, and the comet was recorded from April through to June. Zonares, the Greek historian, says that it was 'as large as the full Moon', and at first devoid of a tail, though a tail developed later (there are no reports of the comet having been seen before its perihelion passage near the end of March). But in England it caused considerable alarm. Harold, last of the Saxon kings, was faced with the threat of an invasion from Normandy, and it seems that the Saxons regarded the comet as an unfavourable omen, though admittedly it had disappeared before Harold met his death at Hastings in October. The picture of the comet is shown in a scene in the Bayeux Tapestry, which was probably commissioned either by Queen Matilda, William the Conqueror's wife, or by Bishop Odo of Bayeux. The

***Fig 5.2** Halley's Comet—Bayeux Tapestry.*

tapestry is embroidered on eight narrow strips of coarse linen; it measures 231 ft long by $19\frac{1}{2}$ in wide, and is now in the town hall at Bayeux in Normandy, though in 1982 it was temporarily removed for renovation. The English translation of the attached legend is 'They are in awe of the star' (*Isti Mirant Stella*). King Harold is tottering on his throne, while his courtiers gaze upward in terror.

1145 This was another well-documented return. The Chinese records describe a long tail, and added that the comet was 'pale blue' in colour, though this must be treated with considerable reserve. There is what may well be a representation of the comet in the Eadwine (Canterbury) Psalter, a 12th century English manuscript. At the bottom of one page there is a stylised view of a comet, probably Halley's.

1301 This return was very favourable, and the comet was seen from places as far apart as China and Iceland. The head was brilliant, and the tail long and broad. One man who certainly saw it was the Florentine painter Giotto di Bondone, who was born in 1267 and died in 1337. It may not have been the only bright comet which he saw (there was another in 1299), but it was very spectacular, and the chronicler Giovanni Villani, also a Florentine, wrote that it left 'great trails of fumes' behind. About 1303 Giotto was commissioned to decorate the interior of a chapel belonging to a wealthy Paduan merchant, Enrico Scrovegni, and he painted 38 scenes, one of which has become famous as the 'Adoration of the Magi'. Obviously he had to include the Star of Bethlehem, and he chose to depict it as a blazing comet dominating the sky. There seems every reason to believe that the model for this was Halley's Comet, and therefore that the fresco shows the comet just as Giotto saw it. The head is in the shape of an eight-pointed star, with layers of pigment built up over it to diffuse the image; by now some of the pigment has been lost, revealing the red adhesive by which it was applied to the plaster.

1378 A less favourable return, but the comet was followed from both Europe and China.

1456 During this apparition the comet was well placed, and, as usual, was regarded as an evil sign. Contemporary writers described it as 'terrible, trailing after it a tail

Fig 5.3 Eadwine Psalter, an illuminated-manuscript collection of the Psalms, was copied by the monk Eadwine from the earlier Utrecht Psalter. Judging from Eadwine's dates, stylistic evidence and this representation of a comet at the bottom of a page, the copy was made in or soon after 1145, when Halley's Comet appeared. The legend refers to the radiance of the comet, or 'hairy star', and remarks that comets appear rarely, 'and then as a portent'. Above the comet are three Latis versions of Psalm 5: Hebraicum, Romanum and Gallicanum (Scientific American).

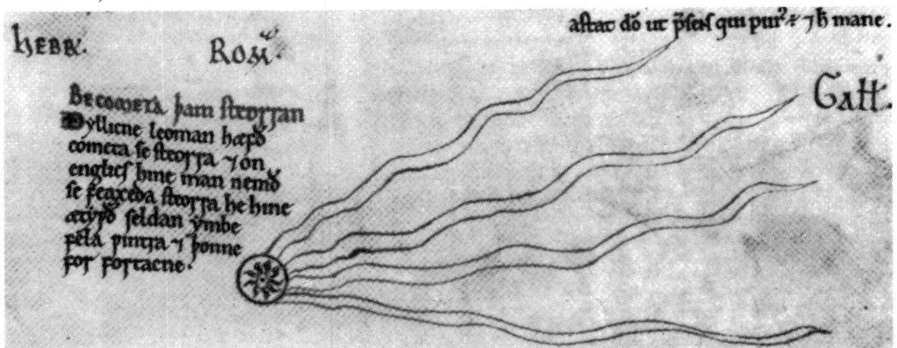

Fig 5.4 *Close-up of comet in the 'Adoration of the Magi' shows how the artist applied tempera and gold pigments to the plastered wall in textured strokes approximating the luminescent appearance of the coma and tail of the actual comet, which he must have observed carefully a few years earlier. He depicted the intensely bright centre of the coma, or head, with what appears to be an eight-pointed star, building up layers of pigment over the star to diffuse the image. Some pigment is lost, revealing the red adhesive by which it was applied to the plaster* (Scientific American).

which covered two celestial signs'—that is to say, 60°. There are also records that the tail was golden in colour, 'at times assuming the appearance of a flame flickering to and fro'.

At that time the Turkish forces were busy laying siege to Belgrade, and it is said that during the night of June 8 a sentry gave the alarm of a fearsome apparition in the sky 'with a long tail like that of a dragon'. The Italian historian and astronomer Pontanus wrote: 'Some persons perceiving the comet in the form of a long sword advancing from the west and approaching the Moon, thought that it presaged that the Christian inhabitants of the West would come to an agreement to march against the Turks, overcoming them. While the Turks, on their part, taking into consideration the state of affairs, fell into no small fears, and entered into serious arguments as to the Will of Allah.'

But if the Turks were disconcerted, so was the current Pope, Calixtus III. It has been said that he went so far as to excommunicate the comet as an agent of the Devil. This may be untrue—it would have been a somewhat pointless exercise—but the Pope certainly did issue a statement, ordering the ringing of church bells 'to aid by their prayers all those engaged in battle with the Turk'.

From Florence, extensive observations of the comet were made by Paolo Toscanelli, who wrote: 'Its head was round and as large as the eye of an ox, and from it issued a tail, fan-shaped like that of a peacock. Its trail was prodigious, for it trailed through one-third of the firmament.' Toscanelli also made some useful positional measurements.

1531 The most important observations during this return were made by Peter Apian, astronomer to the Austrian emperor, who observed from Ingoldstadt in Bohemia and published his results in 1540 in a book, *Astronomicum Caesarium.* He first saw the comet at the beginning of August, and followed it until the end of the first week in September; he observed on every clear night, and tracked it as it moved from Leo through Virgo into Libra. Other reports describe it as reddish or yellowish. It was also observed in China and other countries, and although this was not one of the most favourable returns the comet was certainly conspicuous.

1607 The last return in pre-telescopic times. Among those who observed it was Johannes Kepler, from Prague, who first saw it on September 26 as he was walking home after attending a fête; the head was then of the first magnitude, and there was no tail, though one developed later. About the same time it was seen by the Danish astronomer Christian Longomontanus, from Copenhagen, and it was also described by Sir William Lower from his home in West England. (Lower was later to become one of the earliest observers to use a telescope astronomically, though he was hardly a serious scientist; he described the Moon as looking like a tart that his cook had made!) At its best, the head of the comet seems to have been about the size and brightness of Jupiter, but not quite round, and sometimes described as 'pale and watery', while the tail became long and bright. The comet passed from Ursa Major through Boötes, Serpens and Ophiuchus; an astronomer named Gottfried Wendelin commented that at its best the form was like that of 'a burning lamp' or 'a flaming sword'.

1682 And this brings us up to the return of 1682, which Halley himself saw and which led him to show that the comet was a regular visitor.

Obviously some of the early apparitions have been poorly documented, but they may be enough to answer one important question: Is Halley's Comet becoming less brilliant over the centuries?

That comets are comparatively short-lived members of the solar system is not in doubt. Within the last 200 years several periodical comets have disintegrated; Biela's is the most famous case, but another is Brorsen's Comet, last seen in 1879 at its fifth observed return, and there is also Westphal's Comet, which was particularly interesting in view of its period of 62 years and was on the fringe of naked-eye visibility at its return in 1852. It faded out during the 1913 return, and has not been found since, so that clearly it belongs to the list of comets which have 'gone missing' permanently.

A comet loses some of its material every time it passes through perihelion, and this must be true of Halley's. Presumably this regular wastage will take its toll, and eventually the comet will die, but it cannot really be said that there is any reliable sign of progressive deterioration. True, the comet has never been as brilliant as it was in 837, but that particular return was more or less ideal, and the most important factor seems to be the relative positions of the Earth, Sun and comet near the time of perihelion.

Halley's Comet may be comparatively young, and it is hardly likely that it has made more than about 50 returns, though one can never be sure. However, there is little fear of our saying goodbye to it in the foreseeable future. Astronomers would indeed be sorry to lose it; it has a long history, and we tend to regard it as an old friend.

CHAPTER 6

THE 1835 RETURN

Edmond Halley's great triumph was that he made the first successful prediction of the return of a comet. It was a long time before any more comets were definitely found to be periodical. By 1835 there were two; Encke's, which has a period of only 3.3 years, and Biela's, which was seen in 1826 and came back to perihelion on schedule in 1832 (though, alas, it is no more; it died some time between 1852 and 1865, and nothing now remains of it but tiny meteors). But there was never any serious doubt that Halley's Comet would come back as expected, and preparations to welcome it were made well ahead of time. Mathematical methods had been improved, and there were, of course, the accurate 1759 observations available. Everything indicated that the predictions would be at least approximately correct.

As early as 1817 the Academy of Sciences at Turin offered a prize for the best essay dealing with the movements of the comet. The challenge was promptly taken up by two French noblemen, the Baron Damoiseau and the Count de Pontécoulant, both of whom were distinguished astronomers and both of whom had served in the French Army before turning to science. They made their calculations as accurately as possible, taking all the planetary perturbations into account—including those of Uranus, which had been discovered in 1781 (Neptune, the outermost giant, was not known until 1846). Damoiseau found that the comet would reach perihelion on November 4 1835. Pontécoulant's date was November 13 (or, to be more precise, shortly before midnight on the 12th). Damoiseau finished his work first, and in 1820 the Academy duly awarded him the prize, though in fact Pontécoulant proved to be the more accurate of the two.

What neither Damoiseau nor Pontécoulant had done was to carry their investigations back further than 1759. Otto Rosenberger, a Latvian-born astronomer of German descent, was not satisfied. He had been well trained; for a time he was assistant to the famous F.W. Bessel at the Königsberg Observatory, and had then become a professor at Halle. Rosenberger felt that a new, independent calculation was called for, so he began with the observations of 1682, working out the movements of the comet between then and 1759 and leading up to the expected return of 1835. He found that the four inner planets would combine to speed up the motion of the comet as it neared the Sun, and that perihelion would be correspondingly early; the Earth would account for $15\frac{2}{3}$ days, Venus for $5\frac{1}{3}$ days, and Mercury and Mars together slightly less than

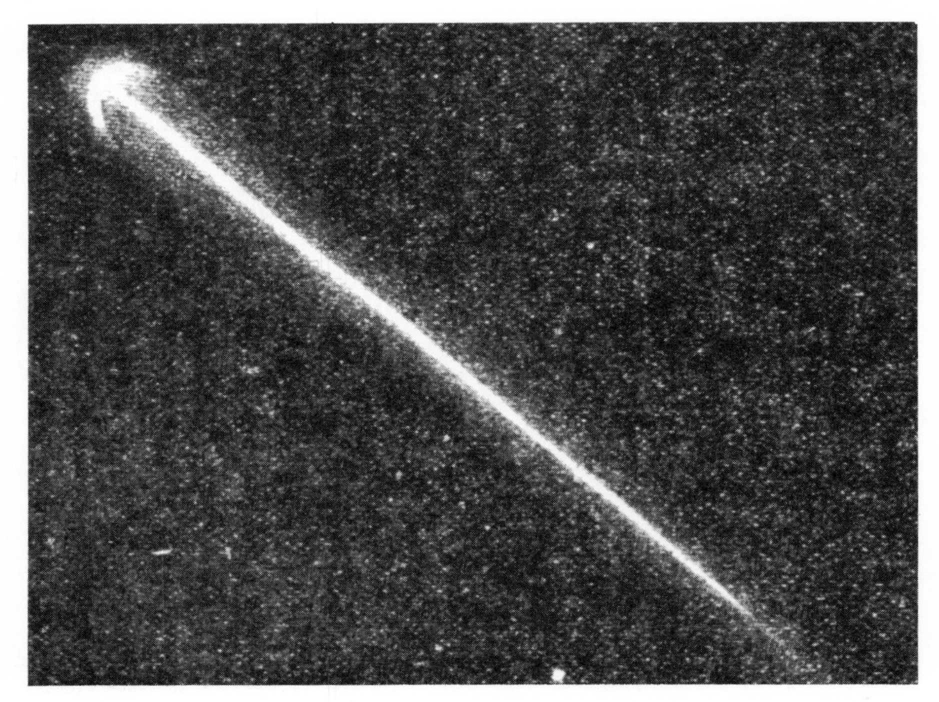

Fig 6.1 *Halley's Comet—as drawn by the French astronomer Arago, October 28 1835.*

one day. Eventually he gave the date of perihelion as November 11. Rosenberger had a competitor in his own country, one Jacob Lehmann, who had taken Holy Orders but had then turned his main attention to astronomy. Lehmann's date for perihelion was November 26.

It was too much to hope that the predictions would be completely accurate, but at least they all agreed that perihelion would occur some time in November 1835, and that earlier in the year it would be in the region of the constellations Auriga and Taurus. Actually, searches began much earlier than that, largely because of the insistence of one of the most celebrated of last-century amateur astronomers, Dr Heinrich Olbers. Olbers knew that it is never safe to predict the magnitude of a comet; Halley's, he maintained, might be brighter than most people expected, and he suggested that the hunt should start at the very beginning of the year or even before. The first attempts were made in December 1834, both in Europe and in South Africa.

At that time John Herschel, son of Sir William Herschel (discoverer of the planet Uranus) had set up a temporary observatory at Feldhausen, some way outside Cape Town, specially to observe the southern stars. Sir William had surveyed the northern hemisphere of the sky, together with that part of the southern hemisphere which is accessible from England, and had discovered hundreds of double stars, clusters and nebulae. He died in 1822; his son determined to complete the survey, and accordingly he went to the Cape, taking with him a fine reflector of 20 ft focal length. He stayed there from 1832 to

1838, and carried through his programme with immense skill and patience. In fact, he followed in Halley's footsteps, but his better equipment and better location gave him far more scope. In his diaries he records that he made his first sweep for the comet on January 29 1835, though without success. He persevered, and continued the hunt whenever possible, but for once he was forestalled. On August 6 Father Dumouchel and Francisco di Vico, at the Collegio Romano Observatory, detected the comet very close to its expected position near the third-magnitude star Zeta Tauri. It was a faint, misty object, seen only with difficulty through the observatory's powerful telescope, and moonlight and bad weather during the next few nights delayed confirmation elsewhere.

Then, on August 21, the comet was seen by Wilhelm Struve at the Dorpat Observatory in Estonia. Struve had the advantage of using a fine refractor constructed by Josef Fraunhofer, then the best lens-maker in the world—and, incidentally, this was the first telescope to be clock-driven (Struve used it to lay the foundations of double-star astronomy). Before long the comet had been sighted by many other observers, and it brightened quickly. Perihelion was reached on November 16, so that of the forecasters Pontécoulant was closest to the truth; but Rosenberger's work was probably the most skilful, and when Struve made his first observation he found that the comet was within 7′ of arc in right ascension and 17′ of arc in declination from the position which Rosenberger had given.

By then the comet had become conspicuous. On September 23 Struve saw it with the naked eye, and a tail was first noted on the following night. On October 14 Struve recorded that the comet was brilliant, with a tail over 20° long. One of the British observers was Admiral W.H. Smyth, who retired from a distinguished naval career in 1830 and set up a private observatory in Bedford. His descriptions of the comet during October are worth quoting:

'October 10. The comet in this evening's examination presented an extraordinary phenomenon. The brush, fan or gleam of light, before mentioned, was clearly perceptible, issuing from the nucleus, which was now about 17 arc seconds in diameter and shooting into the coma; the glances at times being very strong, and of a different aspect from the other parts of the luminosity. On viewing this appearance, it was impossible not to recall the strange drawing of the "luminous sector" which is given by Hevelius, in his *Annus Climeracticus*, as the representation of Halley's Comet in 1682, and which had been considered as a distortion.

'October 11 . . . The tail was increasing in length and brightness, and, which was most remarkable, in the opposite direction to it there proceeded from the coma across the nucleus a luminous band, or lucid sector, more than 60 or 70 arc seconds in length, and about 25 broad, with two obtuse-angled rays, the nucleus being its central point. The light of this singular object was more brilliant than the other parts of the nebulosity, and considerably more so than the tail; it was, therefore, amazingly distinct. On applying as much magnifying power as it would bear, the nucleus appeared to be rather gibbous than perfectly round; but with the strange sector impinging, it was a question of difficulty.'

However, the length of the tail in the pre-perihelion period seemed to alter quite rapidly, almost from one night to the next. Meanwhile Herschel, at the Cape, had made his first observations. In his diary for October 28, he recorded:

Fig 6.2

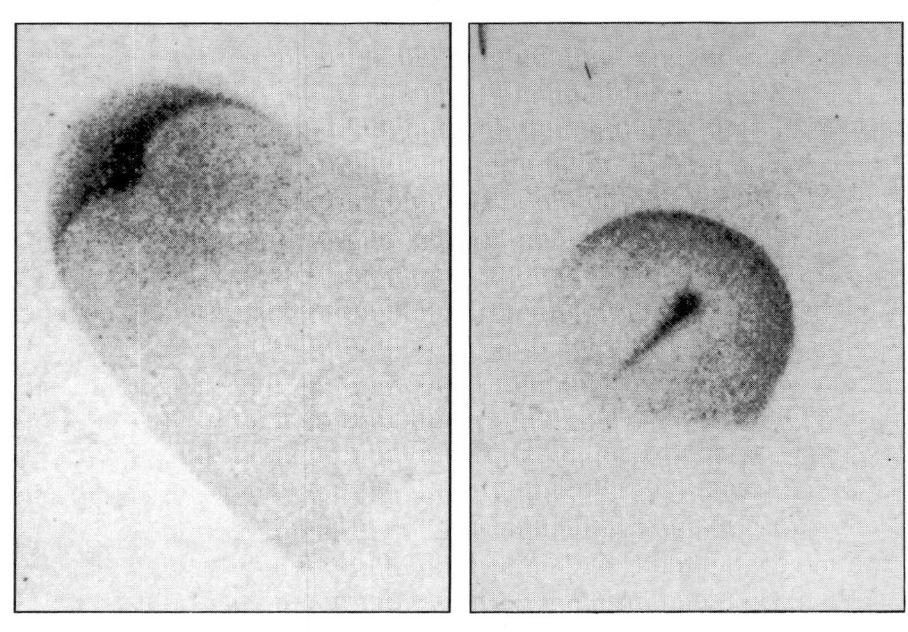

Fig 6.2 *Halley's Comet—Schwabe, October 29 1835.* ***Fig 6.3****—January 26/27 1836.* ***Fig 6.4****—January 28-31 1836.* ***Fig 6.5*** *1836—February 11, May 3.*

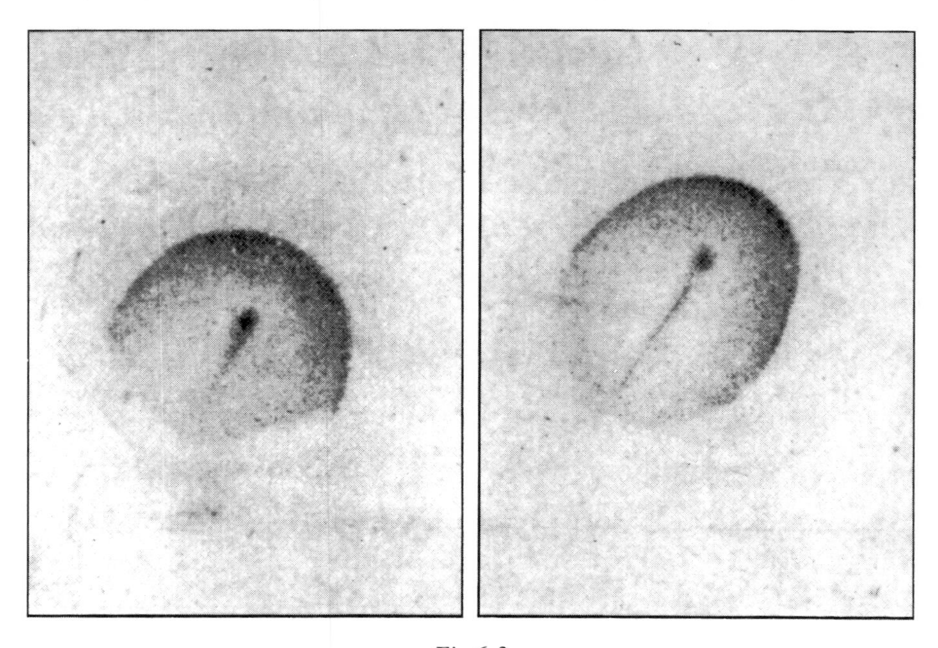

Fig 6.3

Fig 6.4

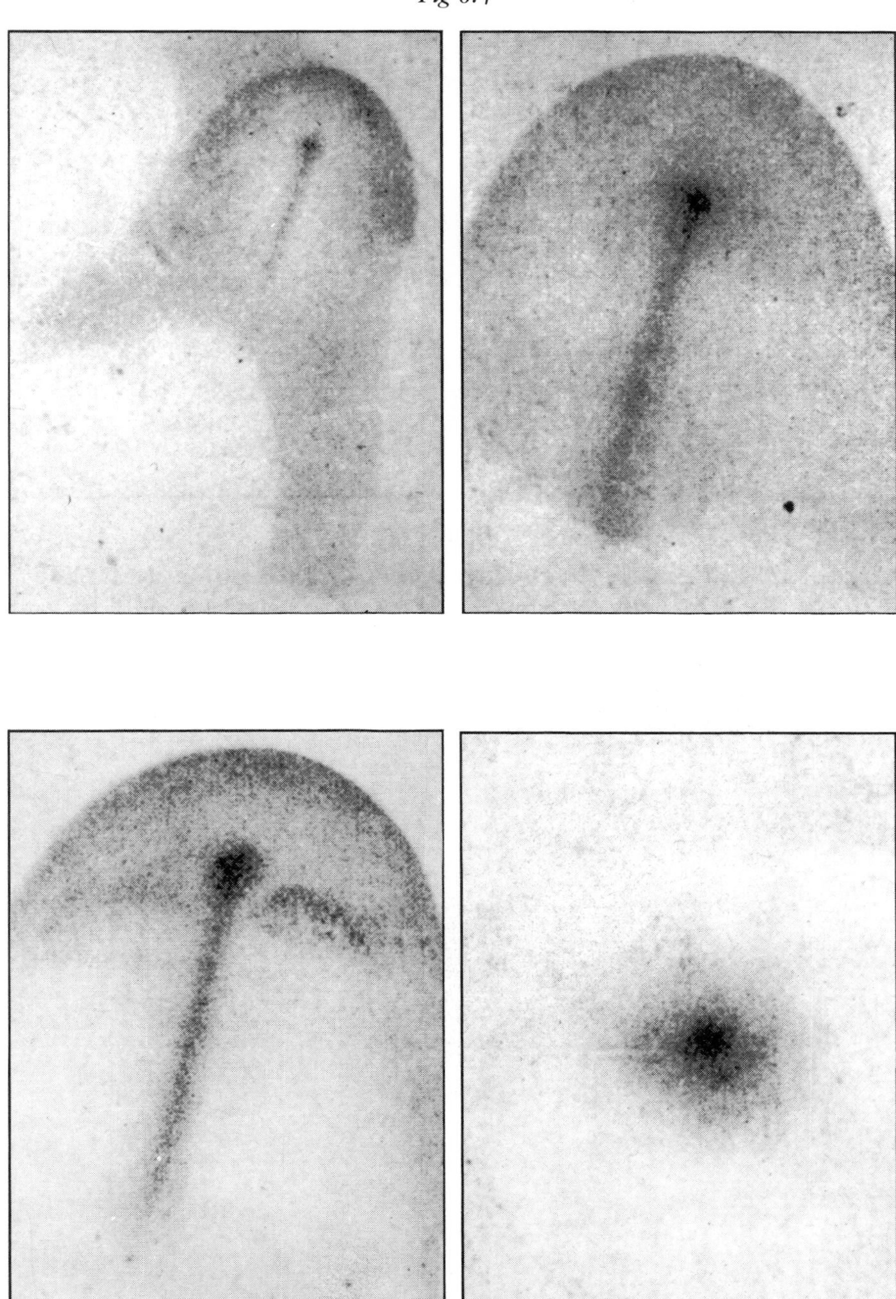

Fig 6.5

'Knocked up a temporary standard for the 7 ft equatorial telescope, dismantled it and carted it out to the first sandhills on the flats; there erected it just at sunset and was rewarded with my first glorious sight of Halley's Comet.' He added that, 'I am sure that I must often have swept with a night-glass over the very spot where it stood in the mornings before sunrise; and never was surprise greater than mine at seeing it riding high in the sky, broadly visible to the naked eye, when pointed out to me by a note from Mr Maclear [Director of the Cape Observatory] who saw it with no less amazement on the 24th'. From Feldhausen, where Herschel's main telescope had been erected, inconvenient trees blocked the sky near the horizon at the critical point. Herschel unceremoniously chopped them down, and on October 29 he duly reobserved the comet.

During October, curious phenomena were seen in the tail. Struve described it as looking like a fan-shaped flame, and subsequently compared it to an oblong, red-hot coal; on October 12 it appeared 'like the stream of fire which issues from the mouth of a cannon at discharge and when the sparks are driven backward by a strong wind . . . at moments the flame was thought to be in motion, or

***Fig 6.6** Feldhausen 1834.*

showing scintillations similar to those of an aurora'. On November 5 the nebulosity itself had a remarkable arched form which was compared with 'a powder horn'.

What could be the cause of these disturbances? Bessel, at Königsberg, concluded that the emission of the tail was of an electrical nature, and went on: 'As the activity increases, the finer particles, repelled by the Sun's light pressure, stream away visibly to form the tail, which grows longer and brighter as the comet approaches perihelion. The emissions from the nucleus sometimes take the form of jets or streamers and sometimes the outflow is more regular, resulting in the formation of envelopes.' We now know that Bessel was on the right track, even though the fact that a comet's tail always points roughly away from the Sun is due mainly to the solar wind rather than to light-pressure.

Perihelion was approaching, and the comet was soon lost in the Sun's rays. The German astronomer Köller, at Kremsmünster, tracked it until November 22, but by that time the tail had contracted—or even, according to some reports, disappeared altogether. After perihelion the comet was seen once more; on December 30 it was picked up by Kreil at Milan, and then by Herschel at the Cape. In a letter to his friend J.C. Stewart, Herschel wrote: 'We have all been staring our eyes out at Halley's Comet, which was so good as to give us a good sight of him in passing to the Sun'. He also commented that the comet

could have been seen during the total solar eclipse of November 20, which was visible from the Indian Ocean area, but there are no reports that the comet was in fact observed.

The disturbances in the tail continued after the comet had become visible again. The best conditions were those in the southern hemisphere, and the observations made at the Cape by Maclear and, independently, by Herschel give us our best information. The tail varied strikingly in length; so did the size of the head, which sometimes looked like a nebulous mass and sometimes almost like a star. Maclear's first sighting was made on January 23, when the nucleus was rather below the second magnitude and there was virtually no tail. 'In the 14 ft reflector it presented an opaque, circular, planetary disk, tolerably well defined, encompassed by a delicately bright coma or halo, which was likewise circular. Crossing the disk in a direction not deviating much from parallelism with the equator, appeared an oblong, elliptical body, distinguished from the rest of the disk by its superior whiteness, and a semblance of greater density. The diameter of the disk measured 2 mins 11 arcsecs; of the coma, 8 mins 12 arcsecs. By the 26th, 'the halo had diminished, and the dimensions of the disk, or body, as it should now be called, were further increased'. A spot like a nucleus could occasionally be seen in the brighter end of the oblong portion. On the 28th, the halo or coma had vanished. Yet the nucleus was distinct, like a faint small star in the following end of the oblong portion. The dimensions of the body had greatly increased, while the intensity of its light had proportionately diminished. The general outline of the cometary body seemed approximating to a parabolic curve, the preceding end of which might be represented by conceiving the tail, as seen before the perihelion passage, abruptly separated from the head, leaving a serrated or ragged outline. The oblong portion with the nucleus resembled a small comet enclosed in the body of a larger one . . .. Throughout the succeeding three months the coma went on increasing, and the outline finally became so faint as to be lost in the surrounding darkness, leaving a blind, nebulous blotch with a bright centre enveloping the nucleus of variable brightness.' Maclear's last observation was made on May 5.

Herschel's observations followed the same general lines; he too saw the remarkable alterations in nucleus, coma and tail, and in the main he agreed with Bessel's theory. 'I cannot help remarking that the conception of a high degree of electrical excitement in the matter of the tail would satisfy most of the essential conditions of the problem. That the Sun's heat on perihelion does eventually vaporise a portion of the cometic matter there can no longer, I think, be any reasonable doubt. That in such a vaporisation, a separation of the two electricities should be effected, the nucleus becoming (suppose) negative and the tail positive, is in accordance with many physical facts.'

Yet Herschel's enthusiasm for his observations of the comet was not whole-hearted. True, he wrote that, 'I have got a fine series of comet observations and hope to see it long after everybody else has lost sight of it', but he also commented that, 'this comet has been a great interruption to my sweeps, and I hope and fear it will yet be visible another month'. By March 25 he wrote that, 'it is fading rapidly, but the nucleus is still pretty bright'; on March 27 it was 'very faint', and it declined rapidly until it had dropped below naked-eye visibility. The last detailed observation was made on May 5, but in a letter to his aunt Caroline, written in the following October, Herschel recorded that, 'we had a fine view of the

Fig 6.7 *John Herschel.*

comet on his return from the Sun. I followed him till about May 20; and I should have kept him in view for longer but for a naughty nebula not down in any catalogue which came so near to his place and looked so like him that he fairly led me off the scent. To say the truth I am glad he is gone.'

Such were Herschel's final comments. He was certainly the last to see the comet; after he had lost sight of it in mid-May it was seen no more until it came back again at the return of 1910.

It must be remembered that for our information about the 1835 return, and indeed for all those earlier, we depend entirely upon visual observations and sketches. Photography had not been developed (incidentally, how many people know that the term 'photography' was first used by Sir John Herschel?) and although Fraunhofer had examined the spectrum of the Sun and mapped the dark absorption lines, it was out of the question to examine the spectrum of a relatively faint object such as a comet. So far as Halley's Comet was concerned, astronomers had to wait until 1910—and Halley did not disappoint them.

CHAPTER 7

THE RETURN OF 1910

The third predicted return of Halley's Comet was awaited with great interest. Since its previous appearance, in 1835, there had been enormous improvements in the techniques of astronomical observation. In particular, the introduction of reliable photographic plates and the spectroscope, enabled astronomers to obtain more information about Halley's Comet, in 1910, than in all of the previous apparitions. Furthermore, during the 75 years since 1835, a number of unexpected, bright and spectacular, very long period comets had been visible, which had done much to arouse considerable interest in cometary phenomena. The Great Comet of 1843 was visible in daylight, and passed within only 830,000 km of the Sun. The tail was the longest of any comet on record, some 320 million km, over twice the distance of the Earth from the Sun. Donati's Comet of 1858 was probably the most beautiful ever seen, and only three years later, in 1861, Tebbutt's Great Comet became a striking object. This comet was possibly the first for which a photograph was attempted. Warren de la Rue tried to record it, but was unfortunately unsuccessful, due to his primitive equipment. Shortly afterwards, in 1864, the first cometary spectrum was obtained by Giovanni Donati, at Florence. The comet concerned was quite bright, and Donati saw that the spectrum obtained was not merely a reflected solar spectrum but contained emission bands which could be due only to materials within the comet itself. The Great September Comet (Cruls) of 1882, was another magnificent object, and an excellent photograph was obtained by Sir David Gill, in South Africa—the first really good comet picture taken. The succession of bright comets continued in 1887, and by this time, cometary astronomers were becoming more skilled in their techniques. Only two years before the predicted return of Comet Halley, experience with the rather beautiful Comet Morehouse 1908 III (shown in Fig 2.9), proved that a continuous series of photographs was vital for recording the rapid changes which could take place in comets, particularly in their complicated tail structure. With this in mind, many scientific expeditions were organised to observe Halley's Comet under the best possible conditions, in places such as Hawaii, Tenerife in the Canary Islands, and Dairien in Manchuria.

Although the apparition of Halley's Comet in 1910 was a reasonably favourable one, there is no doubt that it was, in a way, a disappointment to astronomers. It did not show the rapid and spectacular changes of Comet Morehouse, nor was it as brilliant as the Great Comet of 1882 II, already mentioned. During its period of maximum brightness after perihelion, it was well seen in the southern hemisphere, but was situated unfavourably low for northern observers and, to cap it all, the full moon appeared at the worst possible time. In addition to this, April and early May 1910, around the time of the comet's perihelion, was a period of rather poor weather in the northern hemisphere. Despite these problems, a vast quantity of observational material, comprising visual observations, many hundreds of photographic plates and spectrograms of the comet were obtained. When this information was collated, numerous interesting facts emerged about the development of cometary phenomena. These have been of great importance in the establishment of recent ideas on cometary activity.

Such was the interest generated before the 1910 apparition that, as early as March 1907, the astronomers P.H. Cowell and A.C.D. Crommelin, at the Royal Greenwich Observatory, had published an ephemeris detailing the expected position and brightness of the comet during the approaching apparition. The orbit of Halley's Comet, during the cycle between 1835 and 1910, is shown in Fig 7.1. The comet passed aphelion in 1872, its furthest distance from the Sun, and by 1908 had passed inside the orbit of the planet Saturn on its inward journey. By the Spring of 1909, the comet swept within the orbit of Jupiter, its velocity increasing all the time. The dramatic change in the orbital velocity of the comet is due to its highly eccentric orbit. At aphelion, where the comet is some 35.3 AU from the Sun, it is travelling at only 0.91 km/sec, but at perihelion (0.587 AU), this velocity has increased to 54.55 km/sec. It was expected that Halley's Comet would first be recovered telescopically during the latter half of 1909, and there was some rivalry between observatories with large telescopes, as to who would be the first to pick up the returning comet, last seen in May 1836 from Feldhausen, South Africa, by Sir John Herschel.

During July 1909, the comet would have been an extremely faint object in the morning sky, but badly placed for northern observers. However, no observatory south of the Equator detected the comet at this time. Thereafter, the comet became

Fig 7.1 The path of Halley's Comet between 1836 and 1909.

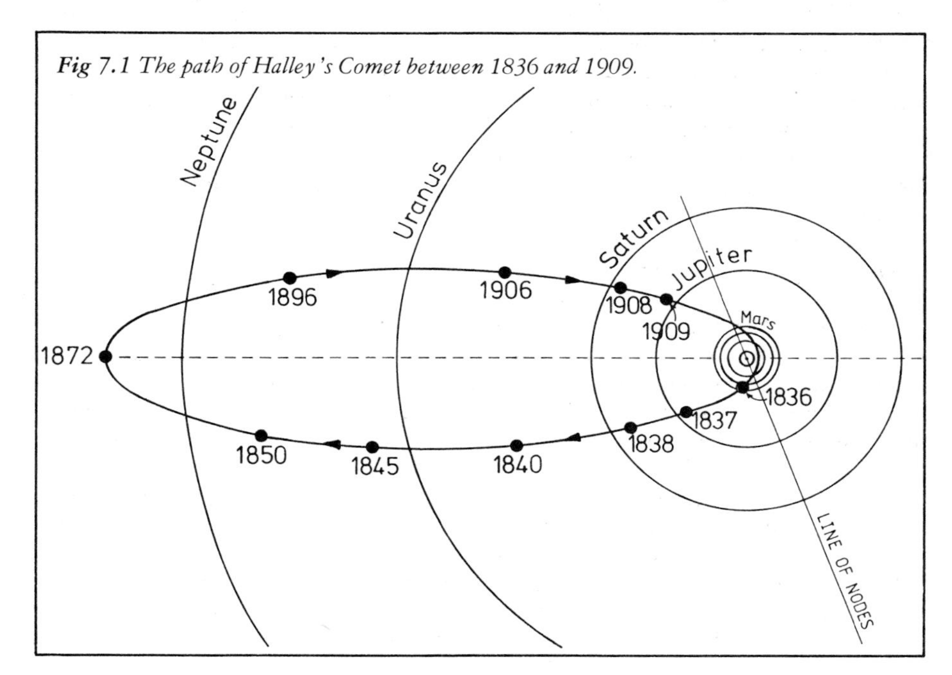

better placed for observers world-wide, and on the night of September 11/12 1909, Professor Max Wolf at Heidelberg, in Germany, first detected the image of Halley's Comet on a photographic plate, taken (most probably) with the Bruce telescope at the Königstuhl Observatory. The comet was found very close to the position predicted in the ephemeris computed by Cowell and Crommelin, as shown below:

Source	**Position of Halley's Comet**	
	RA	**Declination**
Position at $02^h\ 07^m.3$ Königstuhl mean time on September 12 1909	$06^h\ 18^m\ 12^s$	$+17°\ 11'$
Computed from Cowell and Crommelin's Ephemeris	$06^h\ 18^m\ 04^s$	$+17°\ 16'$

Wolf had also suspected an image of the comet on a plate taken on August 28 1909. Furthermore, after the official announcement of the discovery, the comet was found on a plate exposed at Helwan in Egypt on August 24, by Knox-Shaw using the Reynolds reflector. In fact, a photograph of the same region of the sky, on the border between the constellations of Gemini and Orion, obtained at Greenwich, with the 30 in reflector, on the night of September 9/10 also clearly showed the tiny image of the comet. These earlier recorded images were so small and faint that they could not have been positively identified prior to the announcement by Professor Max Wolf. The recorded position demonstrated the remarkable ability and accuracy of prediction of Cowell and Crommelin. As a consequence, they were awarded the Lindemann prize, offered by the Astronomische Gesellschaft, for the most successful prediction. Subsequently, photographs were obtained from the American observatories at Lick and Yerkes. The accurate positions obtained were then used by Crommelin to compute a corrected ephemeris for the comet during the apparition. Calculations showed that Halley's Comet would cross the orbital plane of the Earth, moving from south to north, on about January 14 1910, the Ascending Node of the orbit. Perihelion passage would occur some 96 days later, on April 20 1910, and it would once again cross the Earth's orbital plane (this time from north to south) on May 18 1910. On this latter date, at the Descending Node, the comet would lie 0.15 AU inside the Earth's orbit, and exactly in line between the Earth and the Sun. This event was of major importance, because on this occasion it was predicted that there would be not only a transit of the comet's head across the Sun, but also a possibility that the Earth would pass through the tail of the comet. Unfortunately, only observers in the western United States would have been able to observe the transit, but such a rare occurrence was awaited with great anticipation. This event will be discussed in detail later.

Returning once more to the autumn and early winter of 1909, it seemed remarkable to many people that, although the comet had been discovered in September, it had not rendered itself visible to the unaided eye by the end of the year. Following the discovery of the comet, there had been much excitement, and it was eagerly awaited by everyone. Unfortunately, when first detected in September, it was only visible with a powerful telescope, and was situated at over twice the distance of Mars from the Sun, and considerably remote from the Earth. A diagram showing the motion of Halley's Comet between December 1909 and August, 1910 is given in Fig 7.2. During the latter part of 1909, the distance between the comet and Earth decreased rapidly, but shortly before the end of the year the Earth crossed the imaginary line joining the comet to the Sun, and by early January 1910, the Earth was moving almost straight away from it, as the diagram shows. The rapid brightening and increase in size of the coma will be apparent from the table below, which summarises a few of the observations made up to the end of 1909:

All dates are in 1909

September 11/12 Discovery on a photographic plate by Professor Max Wolf at Heidelberg.

September 16 Observed by E.E. Barnard at magnitude 15.5 with coma diameter of 7″.2 (7.2 arcseconds).

September 26 Barnard found the comet brighter at magnitude 14.5, with a measured coma diameter of 9″.5. Fluctuations in both brightness and diameter were observed from night to night.

October 6 Comet reached 3.0 AU from the Sun.

October 12 Javelle, at Nice Observatory, observed the comet as approximately round, diameter about 10″, with a central condensation of 15th magnitude.

November 14 Magnitude estimated as 13.0 with a coma diameter of 11″.6. The comet was round with no sign of a tail.

November 18 Comet observed at Oxford as a nebulous haze, about 12″ in diameter. Very slight elongation of the coma noticed.

November 30 Barnard measured diameter of comet as 41″·

December 6 Comet reached 2.4 AU from the Sun. In spite of this distance, the comet showed considerable activity, with frequent changes in brightness such as the decided brightness increase of November 22.

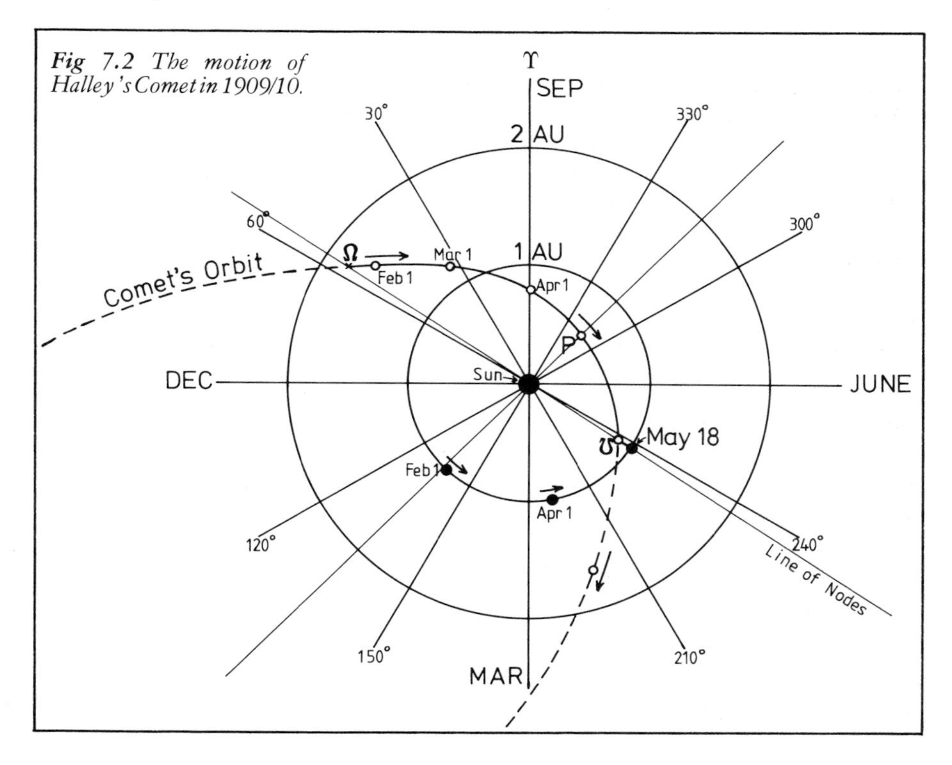

Fig 7.2 The motion of Halley's Comet in 1909/10.

December 13 Photograph obtained with the Crossley reflector, which was the first to show traces of a tail, some 1′.5 in length. The distance of the comet from the Sun was about 2.3 AU.
December 15 The tail extended about 2′-3′, with the brighter part of the coma over 1′ in diameter.
December 19 First series of spectrograms taken by Parkhurst, at the Yerkes Observatory.
December 29 Photograph by Barnard showed a slender, straight tail of some 10′ in length.
December 31 From Oxford the comet nucleus was recorded as magnitude 11.5 and just visible in a 3-inch finder telescope with a magnification of × 23.

Throughout this period the comet was subject to wide variations in brightness from night to night, but was always a telescopic object. By the end of 1909, the total apparent magnitude of Comet Halley was sufficient for it to be seen with a modern 3-inch refractor, but it was far from the spectacular object which many non-astronomers were expecting at that time. In fact, it was not until mid-February 1910 that Halley's Comet was first detected with the naked eye, and in the brief period before this, while Halley's Comet was still brightening, another unexpected but far more brilliant comet made a brief appearance.

The Great January Comet 1910 I, was discovered on January 13, 1910, by some diamond miners in the Transvaal, South Africa, and passed through perihelion only four days later, at a distance of only 0.13 AU from the Sun. The comet was given a

Fig 7.3 The great daylight comet of 1910—Lowell Observatory—January 27 1910.

parabolic orbit for ease of computation, and hence its period (if any) is unknown, but it must be very long indeed. The first accurate observation was made by Innes at Johannesburg, when the comet was closest to the Sun. The comet moved rapidly northwards and was very generally observed in Europe after the 20th. It became visible to the naked eye, even with the Sun above the horizon, and so it is often remembered as the Great Daylight Comet of 1910. A picture of this comet is shown in Fig. 7.3. The comet was rather low from Europe at first, but was an impressive object by January 30. At this time, the head of the comet lay in Aquarius, with the tail arching upwards, some 50° in length, into the constellation of Pegasus. Part of the tail was observed to deviate at a sharp angle from the remainder. Many observations were made during its brief period of visibility, and there is no doubt that it was a far more striking object than Halley became a few months later. Few observations of the Great Comet were made from Europe after the end of February, and the comet was lost altogether by Southern Hemisphere observers in July 1910. Unfortunately, because the Great Daylight Comet was so outstanding, many uninitiated observers of it believed that they had actually witnessed Halley's Comet, this being the object they had been expecting to see for several months.

As it happened, no sooner had the Great Daylight Comet disappeared from view in Europe, than Halley's Comet itself reached naked eye visibility. Once again, it was Professor Wolf at Heidelberg who first saw the comet, with unaided eye, on February 11. Although the comet continued to brighten during late February and March, it became a more difficult object to observe, because at the end of March, the Earth and Halley's Comet were on exactly opposite sides of the Sun. Observations were, therefore, very difficult at this time and, as a result, few if any observations were secured between March 12 and April 12 1910. As if this was not bad enough, moonlight badly fogged many photographs taken during the week just prior to the period of invisibility! Astronomers call this 'Spode's Law'—namely, if things can go wrong, they will! The comet was a naked eye object in the evening sky during late February and early March, but was not spectacular. It did, however, show remarkable activity in its variation of brightness, emission of jets and motion of matter around the nucleus. The first evidence of a broad, stubby, fan-like tail was noted in early February, when the comet was about 1.5 AU from the Sun.

The first observations of the comet after its emergence from the Sun's rays were made around April 12 when the comet's total apparent magnitude was about 2.5, a noticeable naked eye object. The comet reached perihelion on April 20, and much activity in the coma and tail was observed around this time. From mid-April until mid-May 1910, Halley's Comet presented a fine sight in the eastern twilight sky before dawn, being a first to second magnitude object at its best. The planet Venus was also visible in the morning sky at this time, and for a while the two objects appeared quite close together in the constellation of Pisces. The best photographic observations during this period were obtained from observatories on Mount Hamilton, and in Hawaii, Johannesburg, and Santiago in Chile. Comparison of photographs taken on different days, clearly showed emission from the nucleus, the production of haloes, and the formation of jets and streamers. Haloes were observed expanding from the nucleus at velocities between 0.2 and 0.7 km/sec. Considerable activity was also apparent in the tail, with streamers and condensations being observed.

A remarkable phenomenon developed in the tail on April 21. The northern branch of the tail became twisted and knotted, as though some form of sudden

Background ***Fig*** *7.4 Halley's Comet, 30-inch spec, May 10 1910, Helwan, 7-minute exposure.*

Above ***Fig*** *7.5 Halley's Comet, 30-inch spec, June 2 1910, Helwan, 30-minute exposure.*

Below ***Fig*** *7.6 Halley's Comet, 30-inch spec, June 4 1910, Helwan, 30-minute exposure.*

explosion had taken place. The northern edge of the tail, smooth at April $21^{d}.480$*, became jagged and contorted by April $21^{d}.907$, only a few hours later. A great number of secondary centres of condensation were seen, and it appeared as if matter in part of the tail was being expelled in all directions. Another interesting feature of the first week after perihelion was the existence of a persistent jet, apparently supplying material for the northern branch of the tail. It was very bright, but of varying intensity, and some irregular oscillations may have occurred. This jet was noted by many observers, and when it ceased, the southern part of the tail became the brighter. The continual development of streamers in the tail was another feature of this period. Many observations showed evidence of rapid motion in the fine streamers over short time intervals. By the time a fortnight had passed after perihelion, the two types of cometary tail were clearly evident. The gas or plasma tail was sharp and straight, consisting of a number of slender streamers. The dust tail was typically of a diffuse nature, and more strongly curved and fainter than the gas tail, which is normal. Detailed photographs of the comet were taken to reveal the motion of matter in the tail and near the head of the comet. Both photographs and drawings provided a wealth of information, and the continual activity of the comet was evident.

As mid-May approached, the relative motion of the comet and the Earth caused the elongation of the comet from the Sun to decrease once again. During the same period, the path of the comet brought it steadily closer to the Earth. On May 18 it passed between the Earth and the Sun, and was in transit across the face of the latter. It was also predicted that at about this time, if the comet's tail were reasonably straight and more than 24 million km in length, the Earth would actually pass right through the tail of the comet. The transit of a comet across the solar disc is a rather rare occurrence, and it undoubtedly took place on the predicted date—May 18. Astronomers were hoping that if the nucleus of the comet was solid and over 100 km in diameter, it might be possible, using powerful instruments, to see it as a dark speck silhouetted against the Sun, during the course of the transit. Several expeditions were despatched for the purpose of observing this phenomenon, but absolutely nothing was seen. This complete failure to observe a tiny black speck creeping across the solar disc proved not only the highly tenuous nature of the gas in the comet, but also that the nucleus, if it existed, was very tiny indeed, and we now know that this is certainly the case. A similarly unsuccessful attempt to observe the transit of a comet across the Sun occurred for the Great September Comet of 1882, which although so bright that it was visible close to the Sun in broad daylight with the naked eye, also vanished completely when in transit across the solar disc.

Although the transit of Halley's Comet on May 18 1910 certainly occurred (but unobserved), the matter of whether or not the Earth really passed through the tail of the comet for a few hours, a day or so later, is still the subject of considerable argument. The dust tail of Comet Halley was highly curved at the time of encounter on May 19/20, and this fact, coupled with the angle of 18° between the orbital planes of the comet and Earth, probably prevented the Earth from passing through it. Dr Zdenek Sekanina has estimated that the closest dust tail/Earth encounter was roughly 400,000 km, on May $20^{d}.3$, 1910. The closest approach of the Earth to the head of the comet also occurred on about May 20, when the separation was some 21 million km. This compared to a closest approach of only 8 million km at the previous

* Astronomers sometimes use decimals of a day instead of hours, minutes and seconds. For example, April $21^{d}.480$ is $11^{h}31^{m}$UT on April 21.

apparition, in 1835. During the course of May 18/19, 1910, the comet's tail passed from the east over towards the west, and for a couple of days the curvature of the tail caused it to be visible in both the morning and evening sky simultaneously. On the morning of May 20, a broad band of light stretched along the horizon for a distance between 120° and 160°, and at this time Professor Campbell, at the Lick Observatory, saw the comet in the eastern sky. According to him, the tail was at least 140° long, and showing considerable curvature. Many expeditions were mounted by meteorologists, physicists and astronomers to different parts of the world, to see if any electrical, magnetic or other unusual effects could be detected during this period, which might be attributable to the influence of the comet's tail. Their labours were all in vain, and no such phenomena were recorded.

Despite the inconclusive nature of the above experiments, the possibility that the Earth might be plunged into the tail of Halley's Comet had a marked effect upon some of the planet's inhabitants. Although similar passages in the case of other comets had occurred in 1819 and 1861, no one was the wiser until well afterwards, and although some observers claimed to have noticed auroral glows and meteor displays at the time, their veracity was never really established. Perhaps it was these unsubstantiated reports that sparked off wild tales, circulated in the newspapers of May 1910, that the poisonous gases contained in the comet's tail would have a dramatic effect on the Earth. Certainly, comets do contain unpleasant gases like cyanogen and carbon monoxide, but the material is so rarefied that the Earth would be completely unchanged by their presence, and the composition of the Earth's atmosphere would be unaffected. Nevertheless, newspapers reported that in some cities, people were plugging their doors and windows to keep out the poisonous vapours. One often feels that in the 20th century commonsense would always prevail but, quite obviously this is not the case. Around the same period, an enterprising gentleman was known to be selling 'comet pills', but quite what they contained and what they were supposed to prevent is, unfortunately, not known.

For a few days following the closest approach of Comet Halley to the Earth, it was a spectacular object in the evening sky, shortly after sunset, and being north of the Celestial Equator at this time, was well photographed by the American observatories at Lick and Mount Wilson. The tail was some 20° to 30° in length over this period. The weather was also rather favourable throughout the last week of May, and a fine sequence of exposures was obtained. Despite the occurrence of full moon on May 23, some long exposures were obtained because, as luck would have it, a total lunar eclipse occurred on this date! The impression produced by the simultaneous view of two such exciting phenomena was not soon to be forgotten by those who were fortunate enough to observe it. Due to the fine weather prevailing, an excellent series of photographs was secured into mid-June, and many interesting formations in the head of the comet, and the changing structure of the tail, were closely monitored. In addition, a comprehensive spectroscopic study was undertaken, and many emission bands from the molecules present in the comet were measured. From these, a detailed view of the chemical composition of the comet emerged.

The comet grew rapidly fainter as it receded from the Earth. Full moon occurred on June 22, but Halley's Comet was observed by Barnard shortly afterwards, who noted it at about eighth magnitude, with a central condensation within the coma. Two prominent streamers were observed in the tail around this time. As the comet became fainter, fewer observations of it were obtained, and by November 11 1910, Barnard, observing with a 40-inch telescope, estimated its brightness at 11th

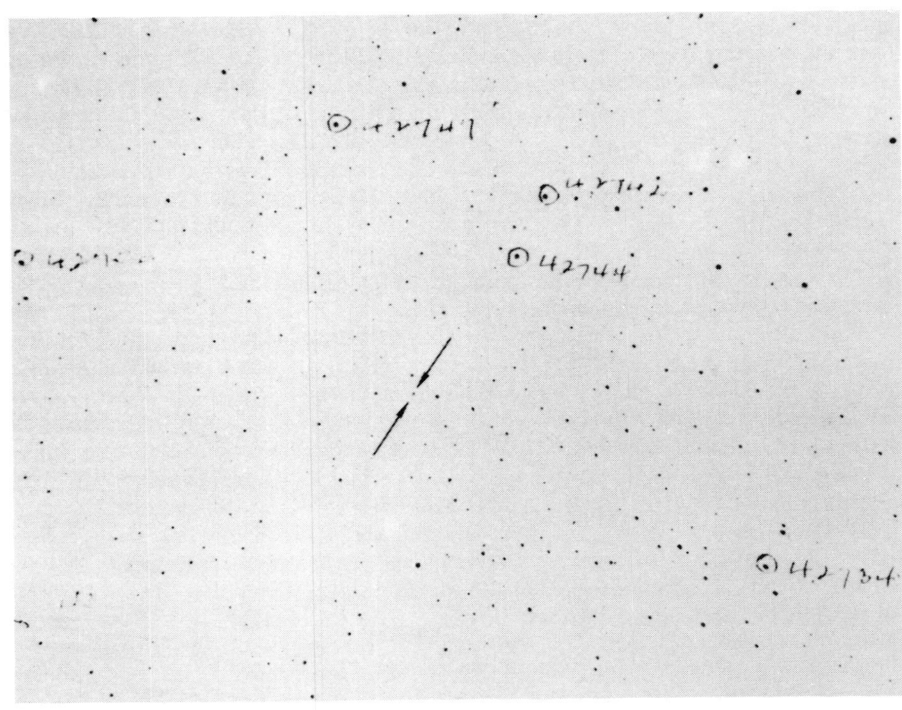

***Fig** 7.7 One of the last photos of Halley's Comet—Lowell Observatory May 30 1911. The distance from the Sun was about the same as that of Jupiter.*

magnitude. By mid-January 1911, it had faded to around 13th magnitude and the diameter of the coma was only 30 arcseconds. A sudden slight increase in brightness occurred at the end of February 1911, the comet brightening by about a factor of two. This persisted certainly until March 4, and possibly longer, by which time the comet was over 4.4 AU from the Sun. During May 1911, the comet faded rapidly, the last visual observation being made by Barnard on May 23, at probably around 16th magnitude. The coma was at least 15 arcseconds in diameter, although it was exceedingly difficult to measure. One of the last photographs taken of Halley's Comet was obtained at the Lowell Observatory, Flagstaff, Arizona on May 30 1911, when the comet was at about Jupiter distance from the Sun. It is shown in Fig 7.7. The last actual photographic observation of the comet showed only a small hazy patch of extreme faintness. It was obtained by H.G. Curtis, on June 15 1911. At this stage, Halley's Comet was well on its way out towards aphelion, a point it reached in 1948, beyond the orbit of the planet Neptune.

CHAPTER 8

METEOR STREAMS AND HALLEY'S COMET

On any night of the year you might perchance look up into a clear, dark sky, and see some meteors (or shooting stars as they are more popularly known). They are, of course, nothing whatsoever to do with stars, but are caused by tiny particles of interplanetary dust (known as meteoroids), being drawn into the upper layers of the Earth's atmosphere, under the influence of its gravitational field. Travelling at tremendous speeds, they are then vapourised in a brief flash of light, by the heat energy resulting from collisions with the air molecules in the atmosphere. The vast majority of meteoroid particles are very small indeed, and these produce meteors which are far too faint to be seen with the unaided eye. However, a meteoroid no bigger than a grape would produce a meteor trail easily visible with the naked eye. This is hardly surprising when one realises that all meteors are travelling at speeds between 11 and 72 km/sec, relative to the Earth at the time of encounter. A typical meteor, such as that depicted in Fig 8.1, will occur at a height of between 80 and 110 km above the Earth's surface, and will normally only last between 0.1 and 0.8 seconds. At least 80 per cent of all the meteors seen in the night sky are what are called 'sporadic' meteors, and these are visible at any time of the year. On any clear, moonless night, a visual observer will see between 5 and 15 of these sporadic meteors every hour. The rate actually varies with the time of day (peak rates occurring at about 4 am) and from month to month, throughout the year.

Most people will also know that, at certain times of the year, the number of meteors visible appears to increase quite noticeably for a short period, causing a phenomenon known as a meteor shower. A meteor shower occurs whenever the Earth passes through a narrow region where the interplanetary dust, which causes meteors, is far more concentrated. These meteor streams, as they are known, result from the large quantities of meteoric dust which are lost by periodic comets, and strewn along their orbits during repeated returns to perihelion. The duration of any meteor shower seen may last from a few hours to several days or even a couple of weeks, depending on the width of the meteor stream. The meteoroids within any particular stream are all moving along parallel paths, but to an observer on the Earth, perspective will make it appear that all the meteors of the shower are apparently diverging from one specific region in the sky. This is called the 'radiant' of the meteor shower. The effect of perspective is also shown by the parallel lanes of the motorway in Fig 8.2 which diverge from the vanishing point on the horizon. It is usual practice to name a meteor stream (or shower) after the constellation in which the radiant lies. Thus, the Perseid meteors, visible every year during the first three

Fig 8.1 Long Perseid streak in Andromeda—August 1977.

weeks of August, radiate from a region in the constellation of Perseus. The radiant point of any meteor stream does not remain fixed in the sky, relative to the background stars, but moves approximately one degree eastwards per day, due to the Earth's motion around the Sun.

The majority of the important meteor showers observed today, are produced by meteor streams associated with short-period comets. A notable exception is the parent comet of the April Lyrid meteor shower (Comet Thatcher 1861 I), which has the longest period of any comet known to be associated with a meteor stream, namely 415 years. Furthermore, two of the currently most active meteor showers, the Quadrantids of January and the Geminids of December, have no known parent comets. However, the orbits of both these meteor streams are of short period, only about 1.6 years for the Geminids, and between 4.7 and 5.2 years for the Quadrantids.

Fig 8.2 Motorway radiant.

There are two meteor streams which appear to be associated with Halley's Comet. One of these occurs in May, the Eta Aquarids, and the other is seen every October, the Orionids. The list below gives details of some meteor streams and their probable parent comets. Several of the usual annual meteor showers are omitted from the list because they have no known cometary association, but the Quadrantids and the Geminids have been included because of their current high activity. In addition, a few of the meteor streams listed no longer produce notable showers, although they have done so in the past, and others are only highly active periodically, with long intervening periods when few,if any, meteors are seen.

Meteor streams and cometary associations

Date of max (approx)	**Stream**	**Parent comet**
Jan 3/4	Quadrantids	No known parent comet
Apr 21/22	April Lyrids	Thatcher 1861 I
Apr 24/25	Pi-Puppids	P/Grigg-Skjellerup
May 5	Eta-Aquarids	P/Halley
Jun 30	June Draconids	P/Pons-Winnecke
Jun 30	Daytime Beta-Taurids	P/Encke
Aug 2	Alpha-Capricornids	P/Honda-Mrkós-Pajdusáková
Aug 11/12	Perseids	P/Swift-Tuttle
Oct 10	October Draconids	P/Giacobini-Zinner
Oct 20/21	Orionids	P/Halley
Nov 4	Taurids	P/Encke
Nov 10-14*	Andromedids	P/Biela (now extinct)
Nov 17	Leonids	P/Tempel-Tuttle
Dec 13/14	Geminids	No known parent comet
Dec 21/22	Ursids	P/Tuttle

* Date of maximum now very uncertain, as very few members of this stream are now observed.

The fact that meteor showers and comets are associated, does not mean that the shower is composed of the débris resulting from complete disintegration of that comet. The April Lyrids, Perseids and Leonids had been recorded for many hundreds of years, although their parent comets were only discovered in the 19th century. There is, however, one example of a meteor shower being enhanced by a complete disintegration of its parent comet, that of Comet P/Biela which divided into two parts in 1846. A drawing by Angelo Secchi of Biela's Comet at this time is shown in Fig 8.3. The two comets formed, returned in 1852, but were never seen again. The meteor shower associated with Biela's Comet was that known as the Andromedids, and it had been seen on several occasions, long before the splitting of the comet. The effect of the disintegration was to cause an enhancement of the shower, due to the introduction of additional débris, and two tremendous Andromedid meteor 'storms' occurred in 1872 and 1885. The evidence accumulated shows that, in general, meteor streams are formed over many consecutive orbits of their parent comet and, therefore, over a long period of time.

The formation of a meteor stream results from the dust dragged out of a cometary nucleus by the continual outward flow of gases. Since a cometary nucleus is most active near perihelion, the greatest quantity of dust will be ejected in that vicinity. The smaller particles form the broad tail of the comet in response to the action of

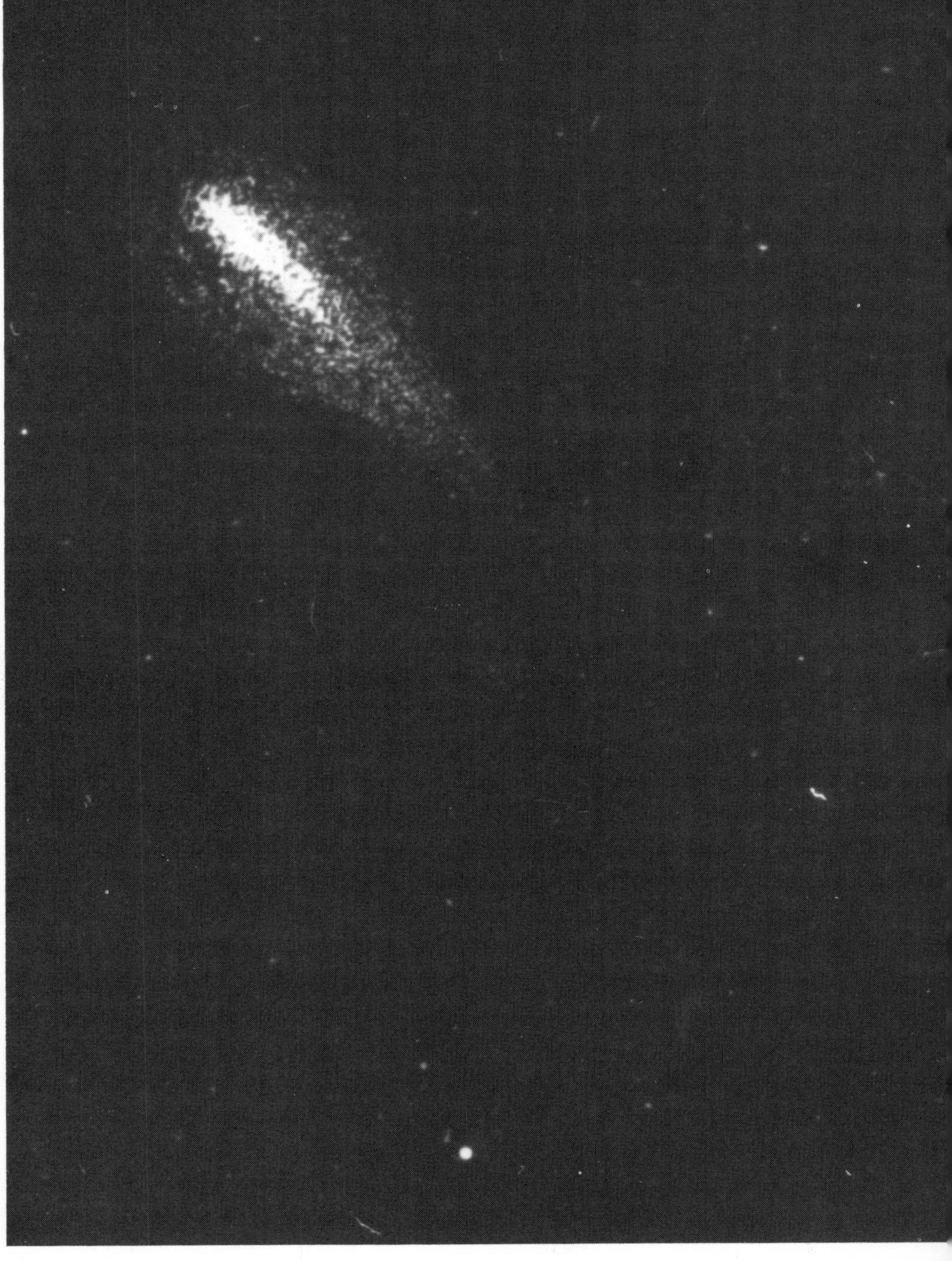

Fig 8.3 Two parts of Biela's comet as drawn by the Italian astronomer Secchi in 1846.

solar radiation pressure. The larger particles, which are relatively unaffected by this influence, go into their own orbits around the Sun. The dust particles which are emitted on the side of the nucleus nearest to the Sun, have orbits of slightly smaller semi-major axis than the parent comet. Their orbital periods will be less than that of the comet and, as a result, they gradually move ahead of the comet nucleus. Conversely, dust particles emitted on the side away from the Sun, have a slightly greater semi-major axis and, consequently, will slowly fall behind the nucleus. The meteor stream therefore consists of dust particles having orbits similar to, but not identical to, the parent comet. The dust particles slowly move further ahead of and behind the comet itself. Eventually, the two groups of dispersing dust particles will pass each other (moving in opposite directions) at the aphelion of the comet's orbit, that is the part furthest from the Sun, as shown in Fig 8.4. At this stage, a complete loop of dust has been formed around the orbit of the comet. The time taken for this to occur is known as the Loop Formation Time and, depending upon the size of the dust particles, diameter of the parent comet's nucleus, and the comet's orbit, may take anything from only 16 years in the case of the December Geminids (very short orbital period), up to several thousand years, in the case of a parent comet of long period.

Although the majority of comets, if not all, produce meteor streams, the particles within that stream will only be observed as meteors in the Earth's atmosphere if the comet's orbit passes closer than a distance of only 0.1 AU from the Earth's orbit and, usually, very much closer than this. In the case of the April Lyrids, the comet's orbit approaches to within only 0.002 AU (300,000 km) of the orbit of the Earth. In general, the closest approach will take place at one of the two nodes of the comet's orbit, where it intersects the ecliptic, or orbital plane of the Earth. The point where the comet's orbit crosses the ecliptic, moving from south to north, is the ascending node, and the point where it crosses from north to south, is the descending node. In a few cases, the closest approach occurs at some distance from the node, but this will only happen if the orbital inclination of the comet is small.

A meteor stream passes through three distinct stages of evolution. Firstly, the clouds of meteoroids ejected from the nucleus extend over a only a small portion of the complete orbit, and are generally concentrated close to the position of the comet itself. In this case, there will be a distinct periodicity in the meteor rates observed, the period being determined by the parent comet. The best example of this type of meteor stream is the Leonids. This shower produced spectacular meteor storms in 1799, 1833 and 1866, in all cases on November 13. An artist's impression of the sight visible in 1833 is shown in Fig 8.5. Here, the clouds of particles are highly concentrated near the comet (Tempel-Tuttle 1866 I), which has a period of 32.9 years. As a result, meteors will only be seen for a few years near the time of the comet's return to perihelion. The shower will also be very short in duration (about four hours), but rates may be extremely high, due to the concentration of dust particles. Unfortunately, planetary perturbations caused the dense, central part of the stream to miss the Earth in 1900 and 1933, but following further perturbations, a tremendous meteor storm was again witnessed at the next return on November 17 1966. At this time peak rates approaching 2,000 meteors per minute were recorded by observers in the USA! Make a note in your diaries to be available on the night of November 17 1999! Another example of this very early stage of meteor stream evolution is the October Draconid stream deriving from Comet P/Giacobini-Zinner.

***Fig 8.4** The formation of a meteor stream.*

Consider two groups of dust particles α and β, emitted from the comet near perihelion, as shown

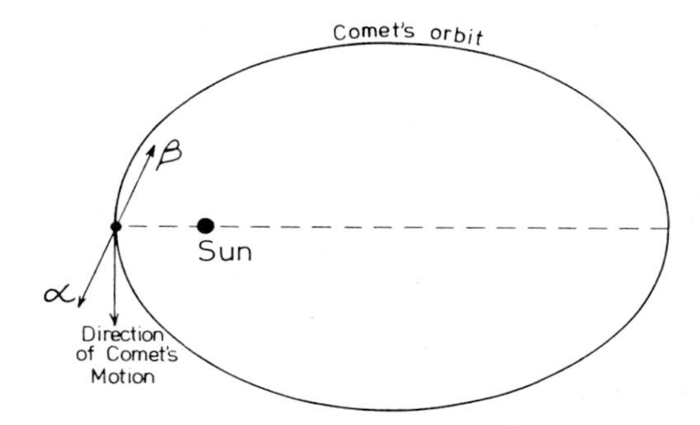

Group α gradually lags further behind the comet.

Group β steadily moves ahead of the comet.

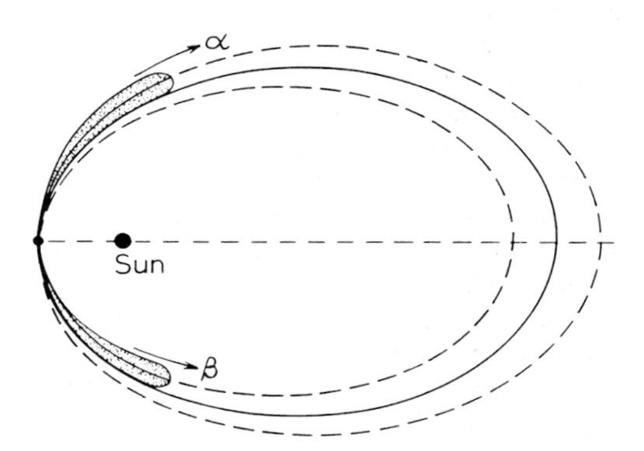

After many revolutions of the parent comet the two groups of dust particles pass each other forming a complete loop of material.

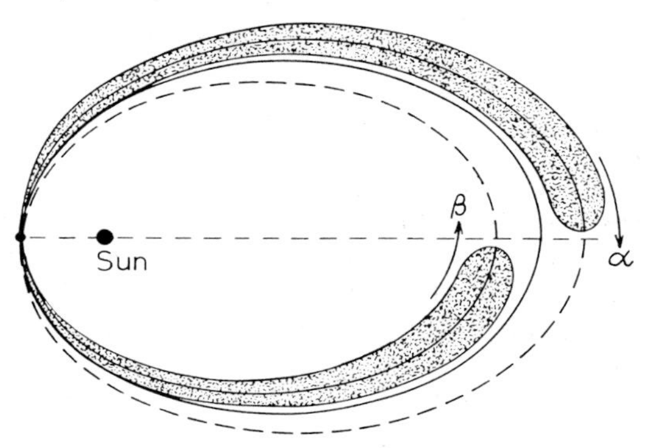

Fig 8.5 Woodcut depicting Leonid Meteor Storm in 1833.

The second stage in the development of a meteor stream occurs as the meteoroid débris spreads out and extends all around the orbit of the parent comet. The meteor stream is still rather narrow, but it will be much broader at the aphelion point of the orbit, than the perihelion. The duration of the meteor shower will still be fairly short, normally no more than a couple of days, but the meteor rates will be reasonably constant from one year to the next, with no long gaps when few meteors are seen. Two examples of this type of meteor stream are the Quadrantids and the April Lyrids. The Quadrantid shower lasts little more than one day, with a very short, sharp maximum, sometimes producing rates of over 100 meteors per hour. The Lyrids may have produced some exceptional displays in the past, but nowadays they generally yield rates of less than 20 meteors per hour, with the shower duration being some four to five days.

The older a meteor stream becomes, the more broadening occurs, due to collisions between the particles in the stream. This occurs mainly near perihelion, where the particle density is highest. Hence, the oldest streams are the widest, resulting in a shower duration of a week or more, and a radiant area in the sky which is generally rather diffuse. Meteor rates may be expected to be very constant from year to year. Two examples of such a stream are the December Geminids and the August Perseids, both of which are showers of constant high activity. The duration of the Perseid shower is currently almost three weeks, so dispersed are the particles within the stream. It is also likely that the two meteor streams probably associated with Halley's Comet, the Eta Aquarids in May and the Orionids in October, are at this rather later stage of stream evolution. The Eta Aquarids have a shower duration of over two weeks, while the Orionids may be seen over a slightly shorter period of about 13 days. Both of these showers will be discussed in detail later. The width of a meteor stream is also dependent on the size of the nucleus of the parent comet, with the larger and more massive comets producing wider meteor streams.

Apart from collisions between the meteoroid particles, other effects cause the steady decay of meteor streams with time. One of the most important influences is something known as the Poynting-Robertson effect, caused by the absorption and subsequent re-emission of sunlight by the small, isolated dust particles within the meteor stream. This process results in the introduction of a very small retarding force on the motion of the particles, which causes a very gradual decrease in the semi-major axis and eccentricity of the orbit of each particle. Ultimately, the particles will spiral in to the Sun, but the effect is only significant for the very smallest meteoroids. For a dust particle having a circular orbit of mean distance 1 AU, a mass of 0.1 gram and density of 0.3 grams per cubic centimetre, the time taken to spiral into the Sun would be 1.3 million years. The Poynting-Robertson effect is dependent on the size of the particle, with the smaller ones spiralling in faster. Hence, within a meteor stream, segregation of the particles will occur according to their size, and the older a meteor stream becomes, a greater proportion of the smaller meteoroids will be lost from the stream. The older streams, therefore, appear to contain a relatively higher proportion of the larger particles (and therefore bright meteors) when compared to the younger streams. As a meteor stream decays, the most probable size of particle gradually decreases, due to collisions between particles within the stream which cause erosion and fragmentation of the larger meteoroids.

A combination of collisional processes within the meteor streams, and the Poynting-Robertson effect, coupled with major perturbations due to close encounters with Jupiter and Saturn, will cause meteoroids to steadily leave the meteor streams and move into independent, randomly distributed orbits in the solar system. These random particles form a general dust cloud which pervades the whole solar system. They are observed as the sporadic meteors, continually incident upon the Earth, in contrast to the stream particles observed only in meteor showers.

The majority of meteoroid particles encountered by the Earth are very small indeed, ranging from 10 microns up to a few hundred microns in diameter, and a mean density of about 0.8 gram per cubic centimetre. A typical shower or sporadic meteor visible to the human eye, has an initial mass of between 0.1 gram and 1 gram, and a diameter of about 1 centimetre. The density of these meteoroids is only about 0.3 gram per cubic centimetre, far less than the density of ice. This low density suggests that meteoroids have a porous, loosely conglomerate structure of particles having a much higher density (say 3 grams per cubic centimetre), but with lots of space in between them, possibly amounting to 75 per cent of the entire volume of the meteoroid. These large spaces were probably filled with ice when the meteoroid was within the parent comet's nucleus.

Recent observations of meteors of all masses, in the range 10^{-10} kg (micro-meteoroids) up to 1,000 kg, using results from artificial satellites, radio meteor detection techniques and visual data, have shown that there is a daily influx of 44 tonnes from the sporadic meteor background. This amounts to a rather staggering 16,000 tonnes per year. Against this background, the mass influx from the meteor streams is rather small, being only about 2.6 tonnes per year for the Quadrantid and Perseid streams, and 15 tonnes per year for the Geminid stream. These observations of the meteor showers provide a means of estimating the total amount of dust in the meteor stream. After various simplifying assumptions, the total amounts of dust in the Quadrantid, Perseid and Geminid streams has been determined as 5×10^9 kg, 2×10^{12} kg, and 9×10^9 kg respectively. These figures represent only a lower limit to the total amount of dust in the parent comet. However, this technique is one of the only ways which we can currently use to estimate the masses of comets directly.

Meteoric particles are being continuously lost from the solar system dust clouds by the combined action of the Poynting-Robertson effect, influx to the major planets, and solar radiation pressure, blowing the very smallest particles out of the solar system altogether. It is estimated that an input of meteoric material to the inter-planetary dust clouds of some 10 to 30 tonnes per second is required to maintain it in a steady state. Many assessments of the total loss rate of all comets have been made, and, of course, great uncertainties are involved, but values as high as 200 tonnes per second (far in excess of the amount required) have been suggested. Even the more conservative estimates give no reason to doubt the hypothesis that comets provide an adequate source of dust particles to maintain equilibrium.

Observations of meteors are invaluable for providing direct information about the composition, density and size distribution of dust particles released from comets. The importance of observing a bright and highly active comet, such as Halley's, has already been emphasised, and it is for this reason that studies of the two meteor streams believed to have been formed by dust emitted from Halley's Comet have assumed such importance.

In its annual motion round the Sun, the Earth samples material from Halley's Comet over two distinct periods, one in May and the other in October. In order to define precisely a point in the Earth's orbit at which any meteor shower has a particular intensity, astronomers use a term known as 'solar longitude'. This is the angle between the Earth-Sun line when the Sun is at the vernal equinox, and the

Earth-Sun line at the other desired position in the Earth's orbit, measured in the direction of the Earth's motion. The solar longitude has a value of 0° on or about March 21-22 each year, when the Sun crosses the celestial equator, moving from south to north.

The descending node of the orbit of Halley's Comet lies at a distance of 0.15 AU inside the Earth's orbit; but due to the fairly low orbital inclination of the comet, the two orbits may approach to only 0.065 AU (9.72 million km) at closest. This point is reached on or about May 8, at a solar longitude of 47°, coincident with the period of visibility of the Eta Aquarid meteor shower, which is active between solar longitudes 39° and 55°. The orbits of the Eta Aquarid particles are not identical to that of Halley's Comet, but the agreement is close enough for little doubt to remain that they are connected with each other. The Aquarid stream has a lower orbital eccentricity than that of Halley's Comet, a semi-major axis of about 5 AU as against 17.941 AU for the comet, and a period of about 11 years, which is of course much less than that of the comet. It is interesting to note that the two orbits are more closely aligned after the comet's perihelion than before (see Fig 8.6), both the comet and the meteoroids having retrograde motion.

The earliest accurate statement that the meteors seen in early May might be connected with Halley's Comet seems to have been made by A.S. Herschel in 1876, although some earlier, rather erroneous statements of this possibility were made by Rudolf Falb in 1868. Unfortunately, although Alexander Herschel showed that the meteors should be visible on or around May 4, and that the radiant point should lie very close to the star Eta Aquarii, he was unable to verify his statement by observation. It was not until May 1866 that W.F. Denning was able to provide the observational evidence required, and that the radiant of the shower seen during the first week in May agreed very closely with that predicted theoretically from the orbit of Halley's Comet. Denning was fortunate in that he had obtained an excellent series

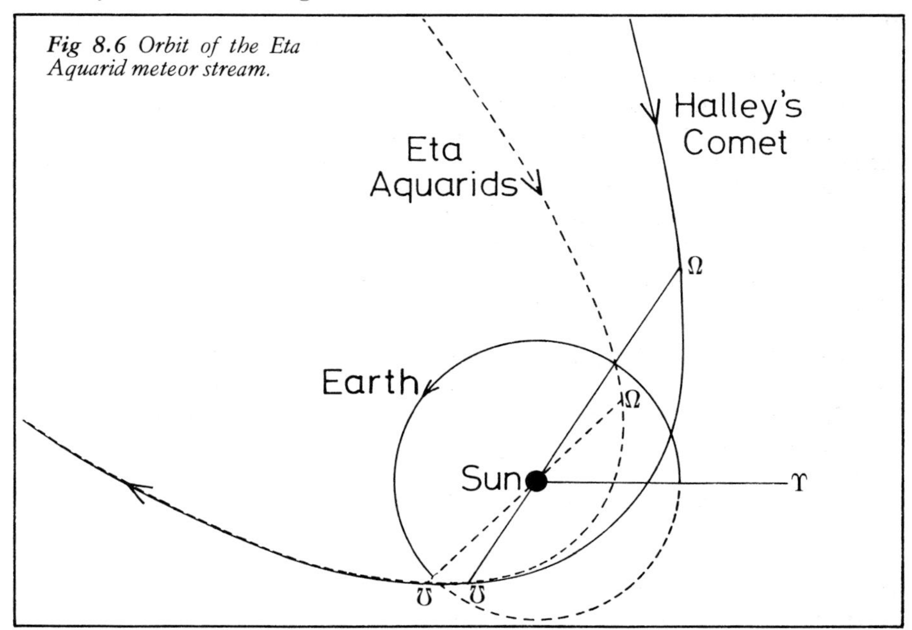

Fig 8.6 Orbit of the Eta Aquarid meteor stream.

of observations of the brilliant display of Aquarids, made by Lieutenant-Colonel Tupman from the Mediterranean on the mornings of May 1 and 3 1870. Tupman's observations, combined with those of some other observers in Italy, showed that the meteor shower was active from at least April 30 until May 6, and that the radiant was very close to Eta Aquarii. Further careful observations were made in 1878 and 1880, confirming the earlier results, and the shower has been reasonably well observed ever since that time.

The radiant of the Eta Aquarid shower is always rather low for observers in the northern hemisphere, and does not attain a respectable altitude before dawn halts visual work, although radar observations can be continued in daylight. This low altitude means that atmospheric absorption renders many of the fainter meteors invisible to a visual observer, and therefore the shower can be well observed visually only from the southern hemisphere. Most of our information about the Eta Aquarids comes from observations made in South Africa, Australia and New Zealand. The activity of the stream was firmly established by the visual observations made by McIntosh, in New Zealand, from 1928 to 1933, and by the comprehensive radio-echo observations made between 1947 and 1952 by Aspinall, Clegg, Hughes and Lovell. More recently, comprehensive analyses of large samples of radar meteor echoes obtained at the Springhill Observatory in Canada and the Ondřejov Observatory in Czechoslovakia have thrown new light upon the complicated structures of the Eta Aquarid and Orionid streams. Moreover, extensive visual observations have been made over the last 20 years by many different meteor groups, world-wide.

The present-day features of the Eta Aquarid stream indicate that it is one of the older meteor streams. Observations of Halley's Comet itself have been made consistently since 240 BC (only one return, that of 164 BC, has been unrecorded), and it is therefore instructive to see whether any very early observations of the Eta Aquarids have also been made. When trying to identify ancient displays of a particular meteor shower, it is important to remember that the calendar date of the shower's observed maximum will not remain constant as time progresses. One reason for this is that the occurrence of annual meteor showers is related to the sidereal year, and not to the tropical year upon which our civil calendar is based. The effect of this is to advance the date of maximum shower activity by one day every 70.59 years. We are, of course, dealing with the familiar phenomenon of precession; its effect upon the date of meteor shower maxima was pointed out by H.A. Newton as long ago as 1863. A second reason for the slow change in the calendar date of maximum is a gradual movement of the orbit of the stream in space, due to perturbations by various massive bodies within the solar system. This causes the orbital node to advance or regress with time. The effect is very small for orbits which are highly inclined to the ecliptic. The node regresses if the stream particles or the comet are moving in the direct sense; if the motion is retrograde, the node advances. Therefore, the retrograde orbit of the Eta Aquarids means that the date of maximum will occur slightly later as the stream grows older.

Careful scrutiny and interpretation of Chinese astronomical records reveals several probable examples of ancient displays of the Eta Aquarids, some of which were apparently very spectacular. The first recorded instance was in 74 BC. The table below lists some early records of the Eta Aquarid shower, together with some more recent results for comparison. In addition to the Gregorian calendar date for the shower, the corrected date, allowing for precession, is given for the year 2000.0.

Year	Gregorian date of shower	Source of information	Equivalent date for equinox of 2000.0
74 BC	April 2	Z. Tian-shan	May 1
401 AD	April 8	Z. Tian-shan	May 1
401	April 9	H.A. Newton	May 2
443	April 9	Z. Tian-shan	May 1
466	April 8	Z. Tian-shan	April 30
530	April 9	Z. Tian-shan	April 30
839	April 13 or 14	Z. Tian-shan	April 29 or 30
839	April 17	H.A. Newton	May 3
927	April 13	Z. Tian-shan	April 28
927	April 17	H.A. Newton	May 2
934	April 13	Z. Tian-shan	April 28
1870	May 3	Lt-Col Tupman	May 5
1878	May 4	Henry Corder	May 6
1930	May 4	R.A. McIntosh	May 5
1933	May 5	C. Hoffmeister	May 6
1970	May 5.5	Brit Astron Assoc	May 6

The dates for the last five entries in the table are those for estimated peak Eta Aquarid activity.

It is interesting to find that of the ancient records (before 1,000 AD), only one, that of 530 AD, occurred in the same year as a perihelion passage of Halley's Comet. On this occasion the shower was undoubtedly spectacular, as shown by the following description: 'Large shooting-stars followed one another north-westwards. Trails which never ceased appearing, numbered in thousands'. Despite this isolated case, there is, however, no real evidence that the Eta Aquarid rates at maximum are significantly enhanced near the time of the comet's return to perihelion.

Observations over the last 50 years give a fairly consistent picture of Eta Aquarid activity during this period. The core of the stream lies between solar longitudes 43° and 47°, extending from the orbit of the comet out to a distance of 0.08 AU. The shower activity varies from year to year, but calculated peak visual rates are generally between 25 and 35 meteors per hour (m/h) for a single observer with the radiant overhead. However, on a few occasions rates as high as 50 m/h have been recorded at maximum. Radar studies have revealed the complicated structure of the stream, there being a considerable variation in particle density along the stream orbit, and also a gradual shift in the solar longitude of peak activity across the stream for consecutive returns.

The activity curve for the shower is not smooth, but shows four or five zones of activity. There is a double peak at solar longitudes of around 44° and 46°.5, subsidiary peaks at longitudes 38° and 50°, and a rather minor 'peak' at longitude 41°. The passage of the Earth through the stream is shown in Fig 8.7. The comet's orbit is at the centre of the diagram, with the x-axis perpendicular to the comet's motion, and the y-axis perpendicular to the orbital plane of the comet. Radial distances from the orbit of the comet are given in AU. The path of the Earth above the comet's orbit is also shown, with solar longitudes marked at 5° intervals. The location of the five peaks in the activity curve are shown. It is clear that if the centre of the meteor stream is assumed to be coincident with the comet's orbit, then the zone of peak meteor activity between solar longitudes 43° and 47° (about May 4 to

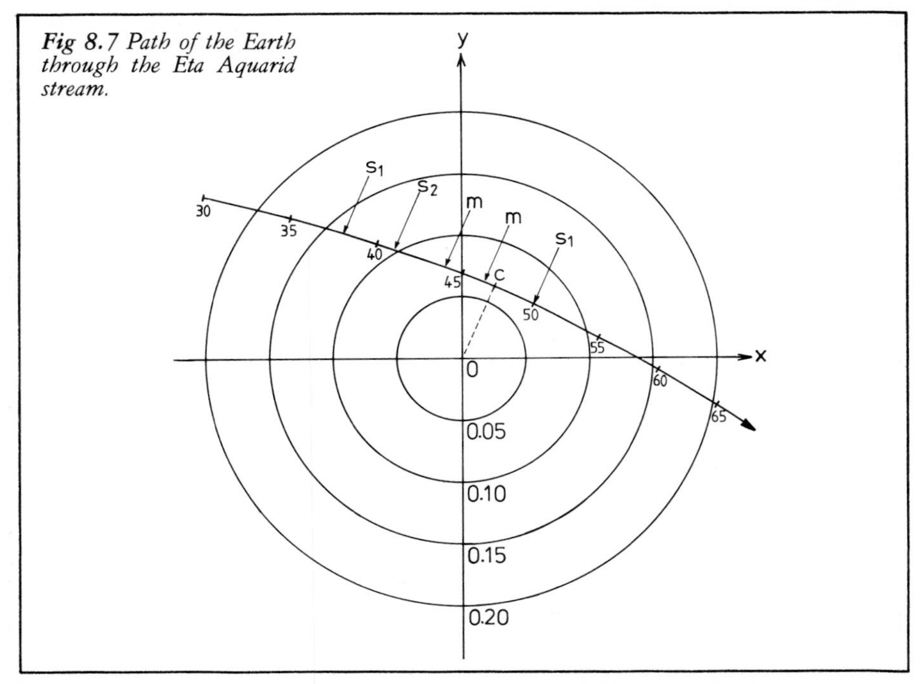

Fig 8.7 *Path of the Earth through the Eta Aquarid stream.*

May 8) does not correspond exactly with the closest approach of the Earth to the orbit of the comet. It is shifted slightly towards the sunward boundary of the meteor stream, so that the Earth encounters the densest portion of the stream before it reaches its minimum distance from the comet's orbit. On average, visual Eta Aquarid meteor rates are greater than 3 m/h between about April 24 and May 20, a shower duration of over three weeks. Aquarid meteors are usually described as fairly slow-moving, with long trails. The radiant of the shower is not a single emitting region, but has multiple centres.

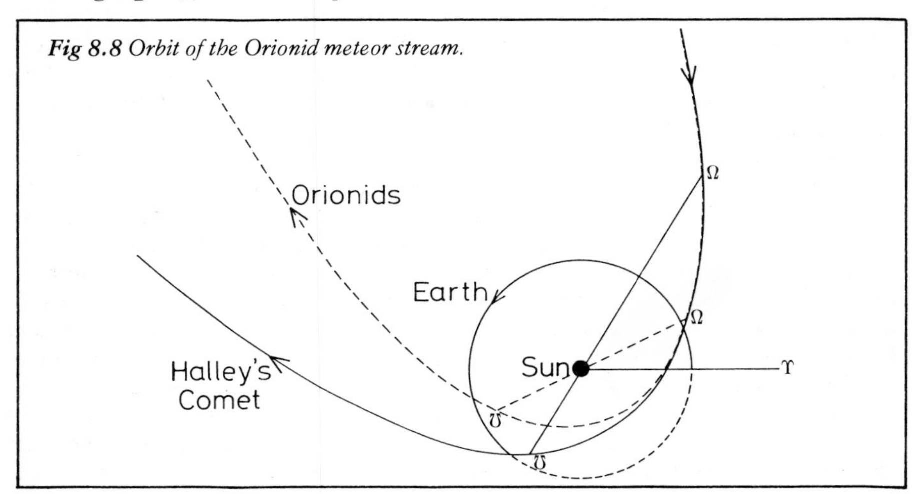

Fig 8.8 *Orbit of the Orionid meteor stream.*

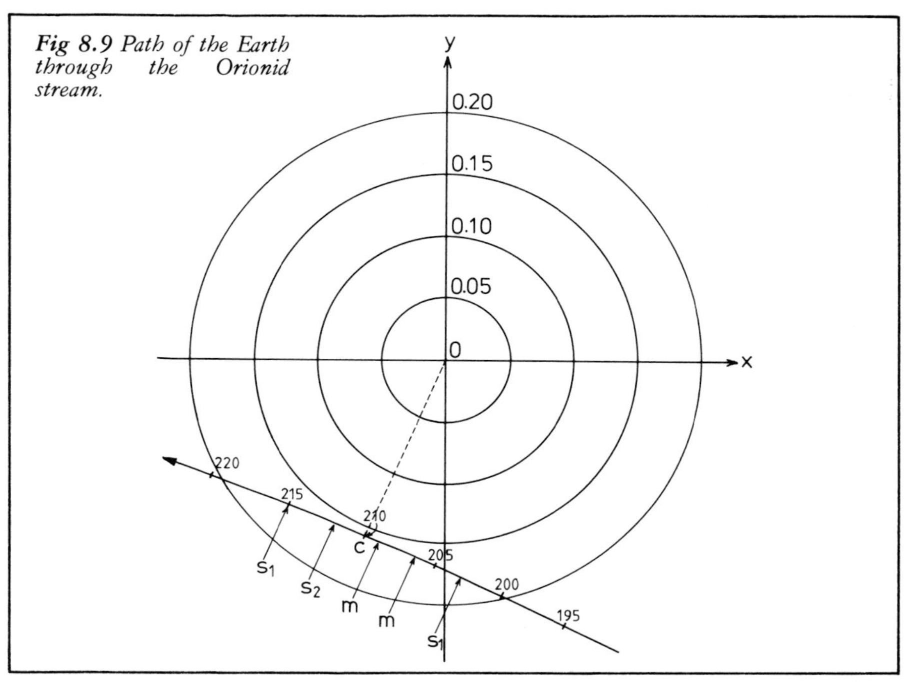

Fig 8.9 *Path of the Earth through the Orionid stream.*

The ascending node of the orbit of Halley's Comet lies 0.8 AU outside the Earth's orbit, but because of the fairly low orbital inclination of the comet the two orbits pass each other at a minimum distance of 0.154 AU (23.04 million km). This point is reached on about October 25, at a solar longitude of 210°, coincident with the period of visibility of the Orionid meteor shower, which is active between solar longitude 203° and 218°. There is great similarity between the orbital elements of the Orionid and Eta Aquarid streams, though both orbital periods are considerably shorter than that of the parent comet. Moreover, the Orionid particle orbits are closely aligned with the orbit of the comet before perihelion passage (see Fig 8.8), whereas the Eta Aquarid orbits are well aligned after perihelion. The general consensus is that, although the Orionid and Eta Aquarid meteors are indeed débris from Halley's Comet, they are probably samples which were released from the comet at decidedly different times in its evolutionary history.

There are only two possible records of any outstanding displays of Orionids in the past. The list compiled by Z. Tian-shan gives 288 AD, September 26 (= 2000 October 20) and 1651 AD, October 14 and 15 (= 2000 October 19 and 20) as times when the shower was very rich. On both occasions it was commented that shooting stars 'fell like rain'. Despite the absence of historical records, the present-day features of the Orionid shower suggest that it also is one of the older showers. The stream is quite broad, activity lasts for about two weeks, and peak rates are reasonably constant from year to year. The Orionids are well-placed for observation from the northern hemisphere, but observations of them have been accumulated world-wide.

In recent years the Orionid stream has been widely studied by a combination of visual, telescopic and radar techniques, and much has been learned about the structure of the stream. As might be expected, there are many similarities between the Orionids and the Eta Aquarids, and a comparison of their properties has yielded much interesting information. The core of the Orionid stream lies between solar longitudes 206° and 210°, but once again considerable variation in particle distribution is observed as the Earth crosses the stream, and there are four or five zones of activity. The double peak occurs at solar longitudes around 206°.5 and 210°, with subsidiary peaks at longitudes 203° and 215°, and a very minor peak at 212°. This complicated activity curve is very like that for the Eta Aquarids. The passage of the Earth across the Orionid stream in October is shown in Fig 8.9. This diagram is similar to Fig 8.7, described earlier, and shows the path of the Earth below the orbit of Halley's Comet, with solar longitudes marked at 5° intervals. The locations of the five peaks of the activity curve are given. Once again a condensed zone of larger particles is found between longitudes around 207° and 209°, but slightly shifted towards the outer, anti-sunward boundary of the stream, and at a distance from the comet's orbit roughly twice as great as for the Eta Aquarids. The Earth encounters the core of the Orionid stream before it makes its closest approach to the orbit of the comet.

The centres of the cores of both the Eta Aquarid and the Orionid streams lie at longitudes of 46° and 208° respectively, and may be connected by a straight line which intersects the orbit of the parent comet, indicating that both zones may well have a common origin. This asymmetric positioning of the two stream cores is probably of evolutionary significance, and may indicate some segregation of stream particles according to their mass—due to the Poynting-Robertson effect. It has been estimated that the formation of a complete loop of meteoroid material around the orbits would need about 15 revolutions. The streams are observed every year for the whole period of revolution of the parent comet, and this sets a lower age-limit for them of about 1,000 years. However, as distinct local irregularities still exist for both streams, the age cannot be greater than 10,000 years, and, of course, Halley's Comet itself has only been reliably observed for just over 2200 years.

The activity of the Orionid stream does vary slightly from year to year, but calculated peak visual rates are generally between 20 and 30 m/h for a single observer with the radiant overhead. The Orionids are active from about October 16 to 30 every year, with peak activity between October 20 and 24. The stream is of added interest because of its multi-radiant structure, which, for many years at the beginning of the 20th century, misled some astronomers into believing that the radiant must be stationary. Following a careful investigation by J.P.M. Prentice between 1928 and 1939, it was shown that the Orionid meteors come principally from three sharply-defined centres (at a mean declination of +15°) at close intervals, and that these centres were in motion parallel to the ecliptic at a rate of approximately 1°.3 per day. These three components were termed the leading stream, mid-stream and following stream. A fourth, more northerly component of the radiant (mean declination +18°) was found to have similar eastward motion, and has been termed the northern stream. This complicated structure is due to the fact that the stream is made up of four or five shells of material, with the particles in each shell having slightly different orbits and peaking in activity at different times. Quite obviously, further research into both the Eta Aquarid and Orionid streams will be essential before a complete picture can be built up. It is particularly interesting that the Earth samples material from the orbit of Halley's Comet from both above and below the comet's orbital plane and at varying distances from the orbit itself—an exceptionally interesting set of circumstances.

CHAPTER 9

THE RETURN OF 1985/6

If the 1910 apparition of Halley's Comet was notable as being the first occasion that photography and spectroscopy were utilised in studies of the comet, then the 1986 return will be equally remarkable in that it represents the first return in the space-age. I am quite sure that if, in 1910, anyone had stated that by the time Halley's Comet returned again man would have flown in space, landed on the Moon, and that close-up scrutiny of the comet from spacecraft was a possibility, they would have been regarded as decidedly 'nutty'! Nevertheless, those are the very realities of the space-age world we live in today. Apart from observations of Halley's Comet from space, one can also be sure that astronomers will no longer restrict themselves to viewing the comet at visual wavelengths, with infra-red and ultra-violet observations now playing such an important rôle. Without doubt, these new techniques will yield still greater quantities of data about this comet than have ever been accumulated before. However, we may live in the space-age but some factors never change. The tremendous interest of ordinary, generally non-scientific people in Halley's Comet will be as intense during the coming apparition as at any time in the past. For many people, the eagerness to view the comet themselves will be a matter of personal satisfaction, irrespective of the scientific results obtained, or new discoveries made about this, the most famous of all comets. It is for such people that this chapter is written.

For any comet which remains in the solar neighbourhood for such a brief period, its visibility depends markedly on the relative positions of the Earth, comet and Sun, during the reasonably short time either side of the perihelion passage, for which the comet is brightest. A comet is at its best near perihelion and so, ideally, we should wish the Earth and the comet to approach as closely as possible to one another around this time. However, there is an additional difficulty, in that at perihelion a comet and the Sun will be relatively close together in the sky and thus observing the comet in a dark sky is impossible. At best, one may observe the comet either in the dawn twilight sky, just before sunrise, or at dusk soon after sunset. In 1910, observers of Halley's Comet were very fortunate in that the Earth and comet passed very close to one another only a month after perihelion, while the comet was still very active. At this time, in mid-May 1910, both the Earth and the comet were on the same side of the Sun. In 1986, a rather unfortunate set of circumstances will render the comet somewhat unimpressive to the casual observer, wherever he may be situated. The comet will be best placed for Northern Hemisphere observers before perihelion, and for those in the Southern Hemisphere, after perihelion. Problems

Fig 9.1 The path of Halley's Comet between 1915 and 1986.

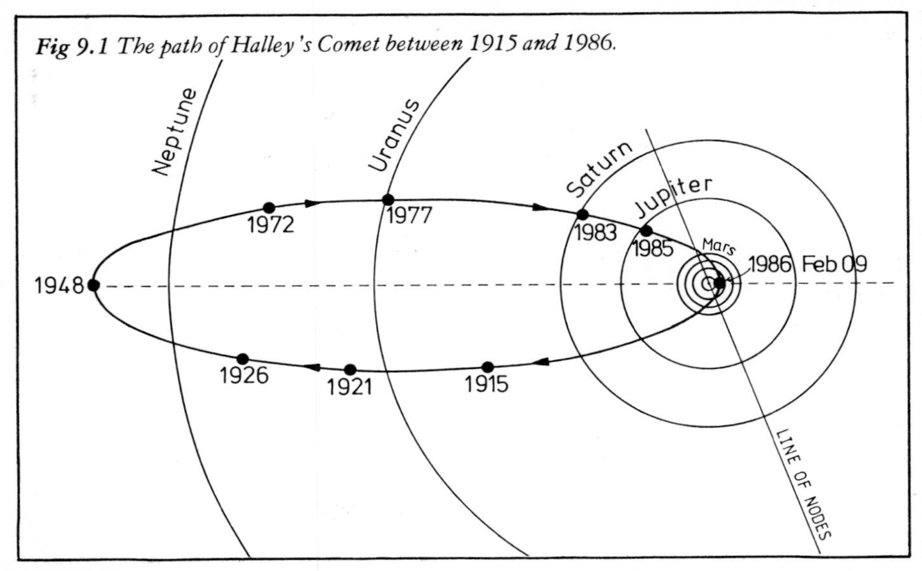

arise because during the first three weeks of February 1986, when the comet is brightest, it will be situated on the opposite side of the Sun to the Earth, and hence will be virtually impossible to observe by ground-based observers. For those with a good pair of binoculars and access to a dark observing site, well away from the glare of city lights, Halley's Comet will be an interesting spectacle from mid-November to the third week of January, and will be at its most impressive in March and early April of 1986. More precise details of the most favourable latitudes from which to observe the comet at any time during the apparition, will be given later.

Comet Halley passed aphelion in 1948, far out from the Sun, beyond the orbit of the planet Neptune, some 35.28 AU from the Sun, and 9.99 AU below the ecliptic plane of the solar system. At this time, it was travelling extremely slowly, at about 0.91 km/sec, and because of this slow movement it spends some 58 years out beyond the orbit of Uranus. The path of the comet within the solar system is shown in Fig 9.1. Once on its inward journey, Halley passed inside the orbit of Uranus in 1977, and that of Saturn in 1983. So, in February 1983, only three years before perihelion, the comet was still some 10.3 AU from the Sun, a distance of over 1,500 million km. As Comet Halley draws in towards perihelion, its orbital velocity increases dramatically, and at perihelion it will be travelling at 54.55 km/sec, about 60 times faster than at aphelion. Not until January 1985 will Halley's Comet pass inside the orbit of the giant planet Jupiter and, by September 1985, it will be sweeping just under the asteroid belt. On November 9 1985, the comet will reach the ascending node of its orbit, at a distance of 1.81 AU from the Sun, and thereafter, for just 122 days, will be above the ecliptic plane. The inclination of the orbit of the comet is evident from Fig 9.2. Its orbital velocity as it passes the ascending node has already risen to 30.52 km/sec, and its distance from the Earth is decreasing. On November 27 1985, only 75 days before perihelion passage, the comet makes its first close approach to the Earth of 0.62 AU. This is far from a good close approach. In 1835, the comet and Earth came to within only 0.05 AU at their closest, prior to perihelion.

Fig 9.2 The orbital tilt of Halley's Comet.

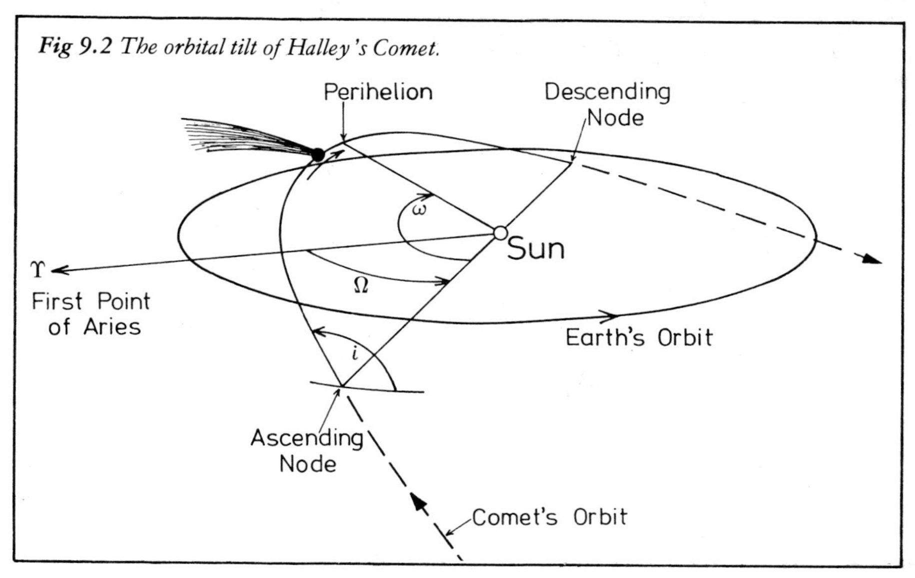

Halley's Comet passes perihelion on February 9 1986, when it will lie at a distance of only 0.587 AU from the Sun, well within the orbit of the planet Venus, but about 0.17 AU above the ecliptic plane. The comet is now travelling at its greatest velocity and swings rapidly past the Sun, to reach the descending node, where it crosses the ecliptic plane once again, on March 10 1986. It was in just this position, back in 1910, that the comet was observed to transit the face of the Sun, and so much excitement was caused by the possibility that the Earth would pass through the comet's tail. Alas, in 1986, the relative positions of the Sun, Earth and comet will be far less favourable. There will, however, be a post-perihelion approach of Comet Halley to the Earth, about 60 days after perihelion, when their separation will be 0.42 AU. This will take place on April 11 1986, and the comet will be very well placed for observers in the Southern Hemisphere at this time. Once again, this distance of approach is a far cry from the 0.14 AU which was their minimum separation in May 1910. The comet is closest to the orbital path of the Earth about 39 days before and after perihelion passage, when it passes above and below the orbit respectively. Unfortunately, this apparition, on these two occasions, namely January 1 and March 21 1986, the Earth is well removed from the locations concerned. The general lack of favourability of the 1985/86 apparition can be attributed to the poor positioning of the comet and the Earth, relative to the Sun at perihelion and also to the rather unfavourable two close approaches, which are not really 'close' at all. The relative positions of the comet and Earth on November 27 1985, February 9 1986 and April 11 1986, are shown in Fig 9.3.

Notwithstanding the far from ideal circumstances of the coming apparition, there will undoubtedly be intense interest shown in the comet by large numbers of people. Quite possibly there will be some who will be disappointed, but for astronomers the scientific rewards of this apparition will be enormous. To help prepare for the 1986 apparition, the National Aeronautics and Space Administration (NASA) in the United States, has organised an International Halley Watch Science Working Group. This body will encourage and coordinate all scientific observations of the comet

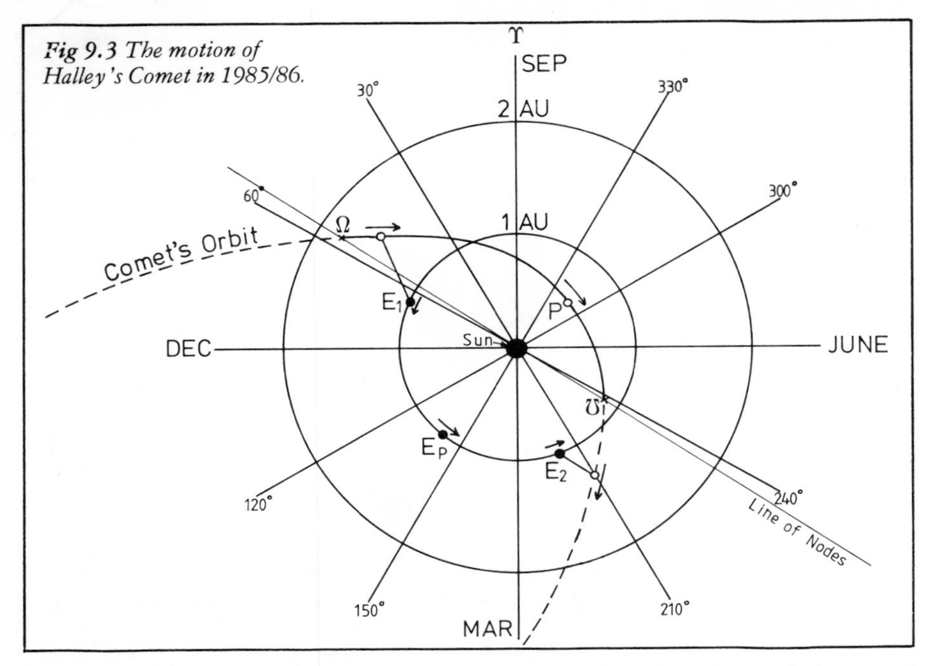

Fig 9.3 The motion of Halley's Comet in 1985/86.

throughout the apparition. It will attempt to standardise observing techniques and instrumentation, and help ensure that all the data and results are carefully collated and properly documented and stored. This is especially important in view of some of the problems which were encountered at the previous return in 1910. At that time, although the project had well defined aims, many observatories did not cooperate with the central committee and, unfortunately, sufficient money and manpower were not available to utilise efficiently the enormous quantity of data which were collected. Indeed, the only comprehensive study dealing with the last apparition appeared in 1931—some 21 years after the comet's perihelion passage!

In 1909/10, all observations were ground-based and made at visual wavelengths. At the coming apparition, it is intended that observations will be carried out not only by space probes sent out to intercept the comet and make observations in its immediate vicinity, but also from satellites in Earth-orbit and possibly by rocket or balloon-borne instrumentation. These techniques, in addition to those employed by ground-based observatories, will cover all regions of the electromagnetic spectrum, from the short wavelength ultra-violet radiation through visual and infra-red bands, up to radio waves. Without doubt, a well co-ordinated effort will be essential to gain the maximum benefit from the comet's short period of visibility. After all, it is a long time until 2061 AD when the next opportunity occurs!

Observations in 1910 have shown that although Comet Halley changes markedly from day to day, it is generally brighter and more active after perihelion than before. Visual magnitude estimates made at the 1910 apparition are shown plotted in Fig 9.4. It will be noted that the intrinsic brightness of the comet naturally increases as its distance from the Sun, r, decreases, but the curve is not symmetrical about the perihelion position. After perihelion, the total magnitude is at first fainter and then substantially brighter than the corresponding magnitudes before perihelion, at the

Fig 9.4 Visual brightness of Halley's Comet in 1909/10.

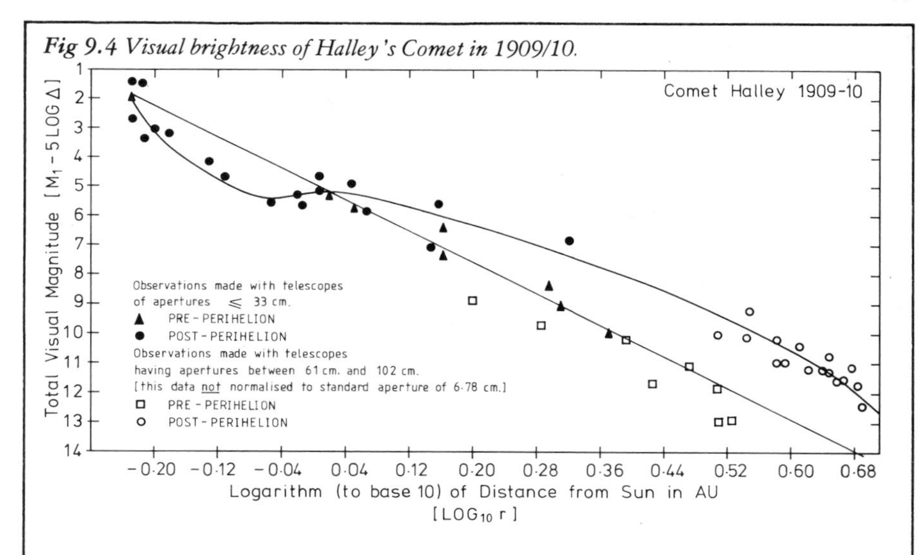

Fig 9.5 Tail length of Halley's Comet in 1759, 1835 and 1910.

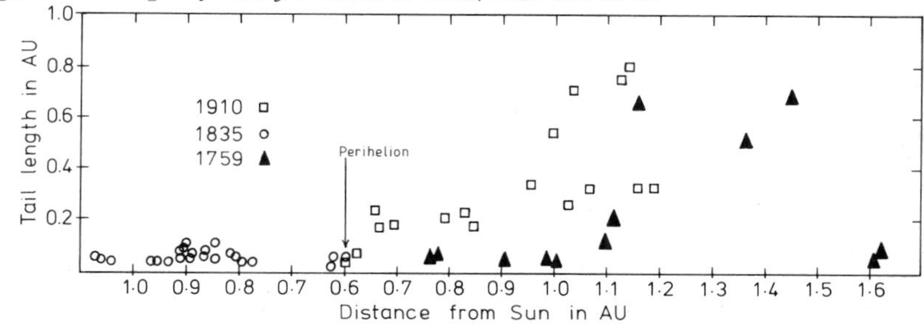

Fig 9.6 Measured coma diameter of Halley's Comet in 1909/10.

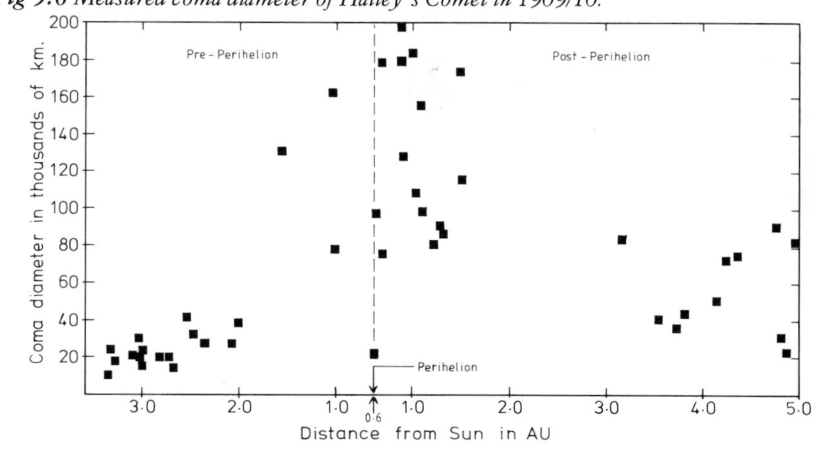

same distance from the Sun. This post-perihelion brightening is well known for Halley's Comet, and various explanations have been proposed for the brightness dip which appears in the curve for, r, between 0.6 and 1.0 AU. Visual magnitude estimates of a comet are very subjective, and depend not only on the observer and type of optical instrument used, but also on the brightness of the background sky. The dip shown in the brightness curve is probably some form of observational effect and not a true picture of the intrinsic activity of the comet.

Other techniques may also be used to assess the activity of the comet. Observations of the tail of Halley's Comet, with the unaided eye, in 1759, 1835 and 1910, have been reduced to show the change in the tail length, with distance of the comet from the Sun. The results are shown in Fig 9.5 and, although there is considerable scatter, the form of the curve is suggestive of the visual tail length being longest after perihelion. A similar analysis of the measured diameters of the coma, made at the 1910 apparition, is shown in Fig 9.6, and once again a slight asymmetry with respect to perihelion is observed. These factors all point to an increased activity after perihelion passage, and it is anticipated that this pattern will be repeated in 1986. With this in mind, the comet will probably be at its most spectacular from mid-February until mid-April 1986.

Apart from the intense interest shown in the comet by professional astronomers, a large number of enthusiastic amateurs and interested members of the general public will be very keen to view the comet as often as possible. Many such people are known to be contemplating expeditions or holidays to parts of the world remote from their homes, to try to find a good vantage point. But where should one go to be able to observe the comet under the best possible conditions? As has already been stated, the comet will be unobservable at perihelion. Furthermore, if as usual the comet is most

Fig 9.7 Most favourable latitude for observation of Halley's Comet.

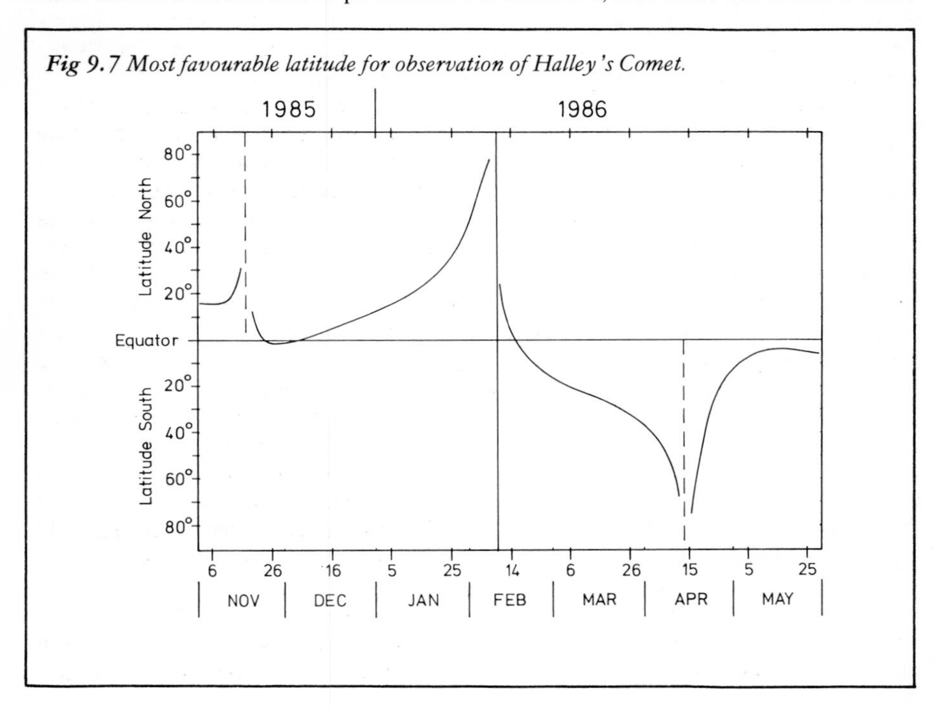

active after perihelion, then those observers in the Southern Hemisphere will get the best view. The comet lies south of the celestial equator from December 24 1985 until long after it has faded from view in mid-1987. However, more important than the position of the comet, north or south of the equator, is whether it is north or south of the Sun. Ideally, we need the comet to have the greatest elevation possible above the horizon at sunrise or sunset. Knowing the relative positions of the comet and the Sun on the celestial sphere enables one to calculate the latitude on the Earth where the comet will have the greatest altitude at sunset or sunrise. In practice, observations are likely to be made either before sunrise or after sunset in a twilight sky. The graph shown in Fig 9.7 gives the most favourable latitude for observations of Comet Halley, for all dates between November 1985 and May 1986. The curve clearly indicates that before perihelion it is best to observe from the Northern Hemisphere and after perihelion from the Southern Hemisphere. If one were to pick two latitudes which would give a reasonable view for the periods before and after perihelion respectively, then one would probably choose latitude 30°N before, and 30°S after. The first of these lines cuts the southern United States, Canary Islands, North Africa, Southern Asia and Hawaii. The second line would include Central South America, South Africa and Australia.

To give intending observers an idea of the circumstances under which the comet may be observed throughout the period November 1985 to May 1986 inclusive, in both the morning and evening sky, a set of drawings, Figs 9.8 to 9.11, is shown, for latitudes 50°N, 40°N, 30°N and 30°S respectively. As has already been stated, the comet will be closest to the Earth on April 11 1986, and the greatest apparent tail length will probably occur some time between about March 20 and April 7. At this time, from latitude 30°S, the comet will be visible in the morning sky, will have an altitude of over 60°, and should undoubtedly be a beautiful sight. From the Northern Hemisphere at this time, it will be much lower and, indeed, from latitude 50°N the comet will never rise at all during the first two weeks of April 1986.

The diagrams show the continually changing altitude of the comet above the local horizon, and its bearing (measured from true north, through east) for both morning and evening apparitions, at either the beginning or end of civil twilight. This is defined as the time when the Sun is 6° below the horizon, either before sunrise or after sunset. This is a rather generous criterion, because normally, astronomers use the start and end of astronomical twilight, when the Sun is 18° below the horizon, and the sky consequently much darker. At these times, approximately 70 to 90 minutes before sunrise or after sunset, the comet would be easier to see against a dark sky, but its altitude would be generally less and, indeed, in some cases would still not have risen (if a morning apparition). Ideally, the positions given in Figs 9.8 to 9.11 should be used as a guide to the location of the comet, roughly 25 to 35 minutes before sunrise or after sunset. However, observers would be well advised to start their searches either about 1½ hours before sunrise, or very soon after sunset, to give them the maximum time to view the comet, either as soon as it rises (morning) or as soon as possible after sunset (evening). In this way, observers should not miss any brief sightings. Bearing in mind that the comet will be a rather faint object, it is suggested that observers choose a site where the skyline is unobstructed in the direction in which the comet will be seen.

Use a pair of binoculars to find the comet in the twilight sky, but under *no* account sweep for the comet with the Sun above the horizon, if there is a danger of the Sun entering the field of view, with disastrous consequences to the observer's eyesight.

Changing altitude of Halley's Comet above the local horizon for four different latitudes, 1985/86.

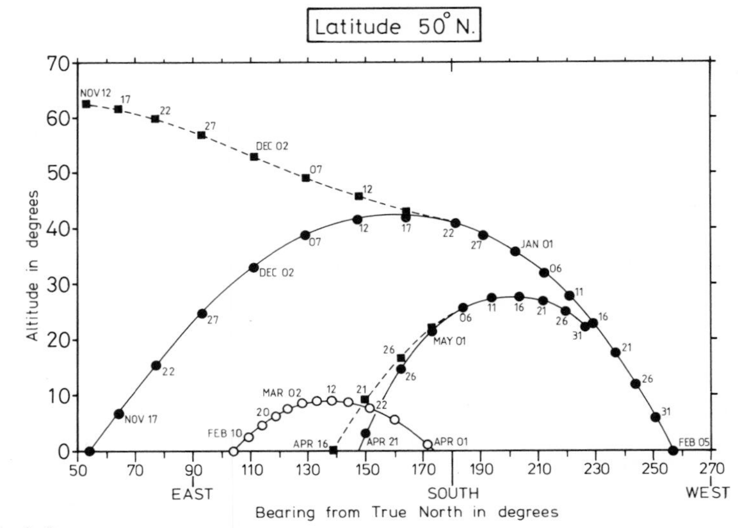

Fig 9.8

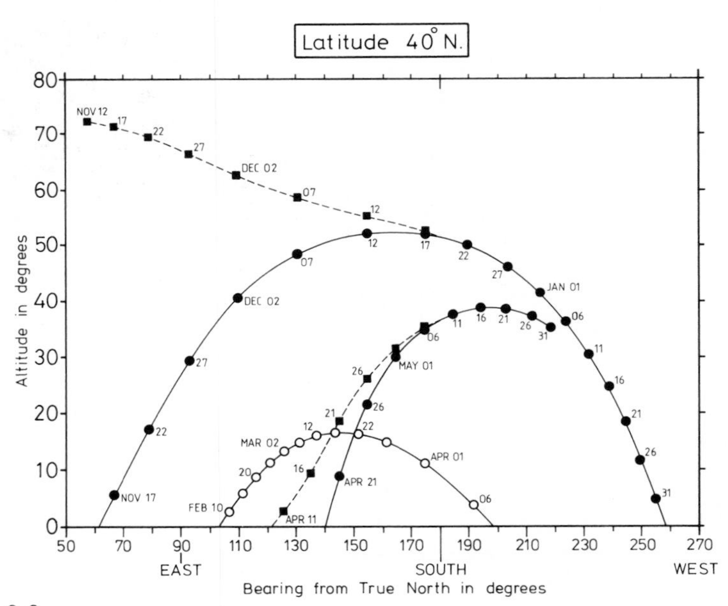

Fig 9.9

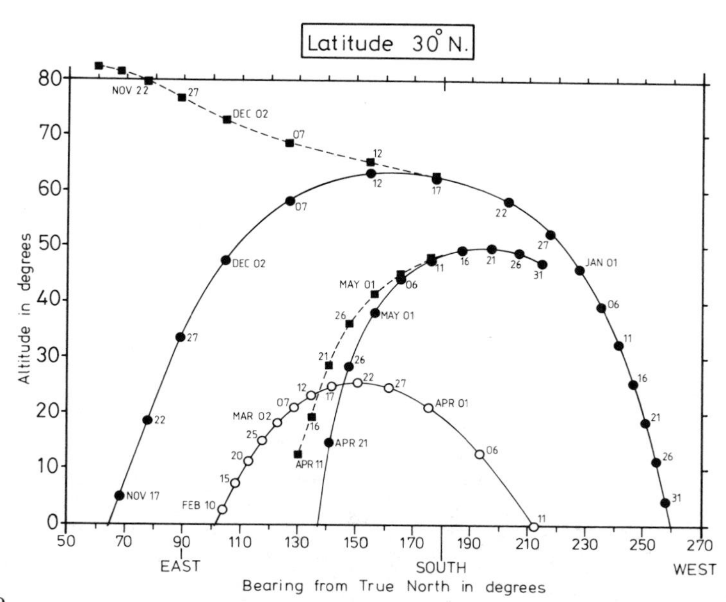

Fig 9.10

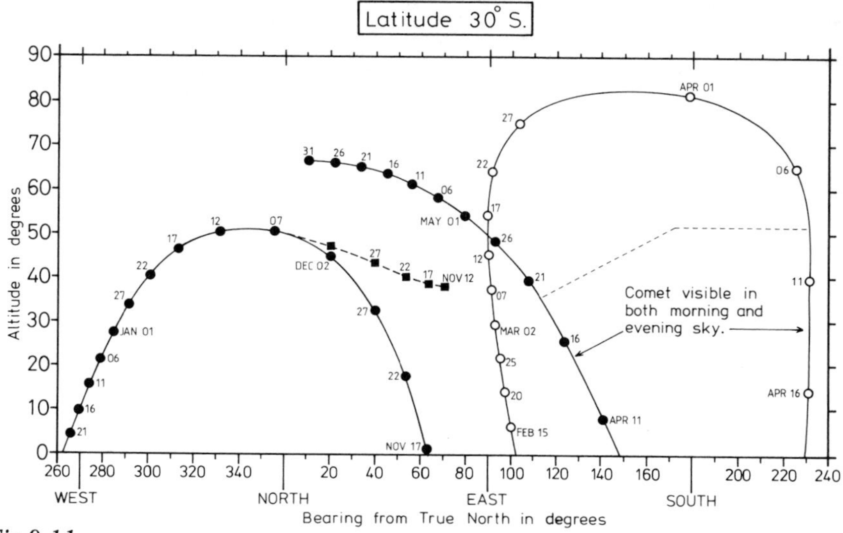

Fig 9.11

Of course, the tips given here only really apply when the comet is rather unfavourably placed for a particular latitude, or is fairly close to the Sun in the sky. It should not be forgotten that from the Northern Hemisphere there are long periods in December 1985 and January 1986, and also late April/early May 1986, when the comet is in the evening sky and still has a respectable altitude well after sunset, when the sky is really dark. In these cases, observation with binoculars or a telescope will provide a superb view, and those keen on photography will no doubt try to record the comet on film. Similarly, from latitudes south of the 20°N line, a fine view will be obtained during March and early April 1986, when the comet is at its most spectacular, and observations are again possible in a dark sky. Over the same period, the comet will be extremely difficult to observe from northern latitudes, due to its very low elevation in the bright dawn sky, and this situation will not improve until after April 21 1986, when the comet will reappear in the evening. In a similar way, Southern Hemisphere observers will have a rather poor view of the comet before perihelion, when it is best placed for those in the north. One other point of interest is that the comet will pass near opposition in April 1986, and consequently from latitudes south of 20°N the comet will be observable in both the evening and morning sky, for a short period in mid-April, 1986.

The greatest tail lengths will probably occur sometime in late March or early April 1986, but another aspect of importance is the apparent total brightness of the comet. Two curves of the expected brightness are shown in Fig 9.12. The curve M_1 gives the predicted total apparent magnitude of the comet as a function of date. The second, lower curve, labelled M_2, refers to probable nuclear magnitudes of the central condensation within the coma. The comet is, of course, brightest at perihelion, when it reaches $M_1 = 2.9$, a third magnitude object and not unduly spectacular. Unfortunately, it is not observable at this time and, by the end of February 1986, has already faded to $M_1 = 4.5$. The horizontal dotted line marked across Fig 9.12 shows the theoretical threshold of naked eye visibility, under ideal conditions with the comet in a dark sky. In theory, the comet should be visible to the naked eye from mid-December 1985 until the end of April 1986, a period of $4\frac{1}{2}$ months. However, it is not likely to be seen consistently throughout this period from any one point on the Earth. The slight bump which occurs on the descending portion of the two curves in Fig 9.12, is due to the close approach of the comet to the Earth during the first week of April. This will undoubtedly be the time when the comet is at its best, alas for northern observers who will be extremely fortunate if they get a view of it at this time.

For the astronomer with a telescope, the comet will be a rewarding object. It should be visible in a 6-inch telescope from mid-September 1985 until mid-August 1986. Members of the general public would be well advised to seek out their local astronomical society or organisation, who will no doubt have telescopes capable of providing an excellent view of the comet's structure against a dark background sky. They will probably be only too willing to enthuse about their hobby to those with a more casual interest. Telescopic observers will be able to watch the progress of the comet as it shifts slowly from constellation to constellation against the background stars. The apparent motion of the comet will be at its most rapid during the first two weeks of April, because it is then at its closest to Earth. However, at no time will the comet be observed to 'streak' across the sky, and its motion will only be detectable through a telescope, or from one night to another.

The path of Halley's Comet against the star background, is shown in the four

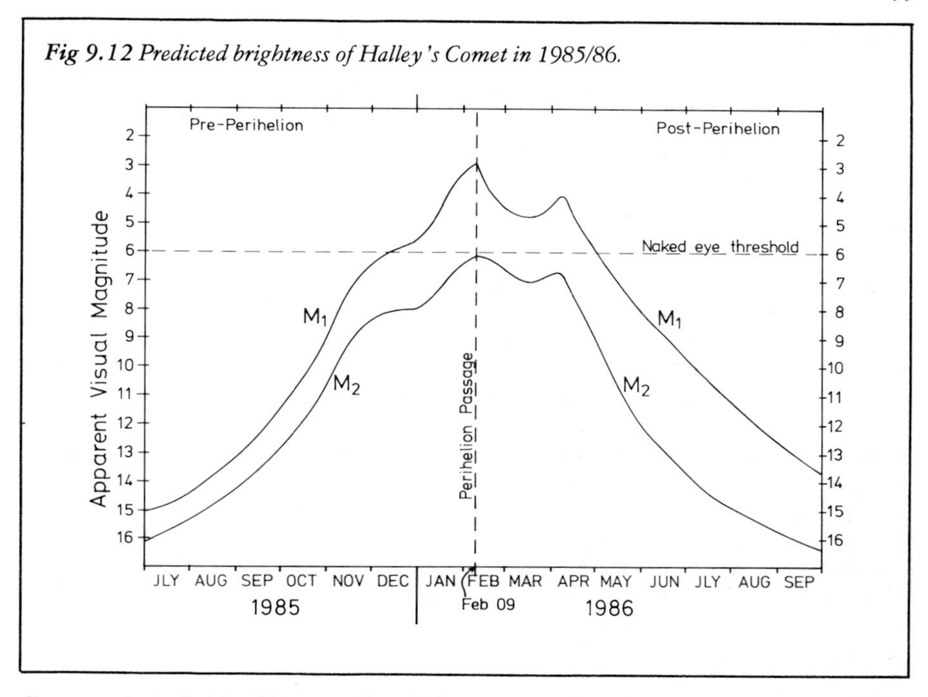

Fig 9.12 *Predicted brightness of Halley's Comet in 1985/86.*

diagrams included in Chapter 12, which comprises the Summary and Appendices. These diagrams give an impression of the movement of the comet from January 1985 until March 1987. Between January and the end of October 1985, the comet's motion will be slow and it will pass alternately between the constellations of Orion and Taurus, above the familiar pattern of the stars of Orion the Hunter. The comet will only be visible in a telescope at this time. The comet passes once again into Taurus on October 24 1985, and thereafter its westward motion becomes more rapid. An interesting opportunity to locate the comet occurs at about 00.45 UT, on November 12 1985. At this time, the comet passes between the close pair of stars, 65 and 67 Tauri, magnitude 4.4 and 5.4 respectively. It will be about eighth magnitude at this time, and visible in a small telescope or powerful pair of binoculars. At closest, the comet will pass about one arc minute south of the star 65 (Kappa) Tauri and roughly five arc minutes north of 67 Tauri. A similar opportunity occurs a few days later, on November 14.57 1985, when the comet passes between the two stars 37 A^1 Tauri and 39 A^2 Tauri, magnitudes 4.5 and 6.0 respectively. On the night of November 16/17, the comet will pass just less than $2\frac{1}{2}°$ south of the Pleiades star cluster in Taurus, popularly known as the 'Seven Sisters'. The comet will be a seventh magnitude object at this time and visible in binoculars. During late November the comet moves through Aries (the Ram) and on November 28.5 will be very close to the Messier object M74 (NGC 628). This is a type Sc spiral galaxy of about the tenth magnitude, and lies slightly east and north of the star Eta Piscium. The comet's total apparent brightness will be some 40 times greater than the galaxy at this time.

During December 1985, the comet moves through Pisces, below the Great Square of Pegasus, and on December 22 1985 passes into Aquarius (the Water Bearer). It will be situated less than a degree south of Eta Aquarii on the night of December 26/27, and of Gamma Aquarii on New Year's Eve. The comet should be a naked eye object by this time, being of the fifth magnitude. The comet remains in Aquarius until perihelion, by which time it will have been lost to view in the evening twilight sky. After perihelion, the comet's apparent daily motion across the sky increases rapidly, moving through Capricornus in early March, Sagittarius and then into Corona Australis by the beginning of April. By this time, the comet will lie far south of the celestial equator, conditions very favourable for those in the Southern Hemisphere. As it draws in to closest approach with the Earth on April 11, it rapidly moves from one constellation to another, passing through Scorpius, Ara, Norma, Lupus, Centaurus and Hydra in the space of just three weeks. By the end of the third week of April, the comet is once again moving northwards, back towards the celestial equator. Its mean daily motion decreases markedly during May 1986 and thereafter, until March 1987, it remains in the vicinity of the constellations Sextans, Crater and Hydra.

Apart from the comet itself, ground-based observers will also be keen to monitor the activity of the Orionid and Eta Aquarid meteor showers, believed to be debris from Comet Halley, during the 1985/86 return. Both showers are annual events, but in October 1985 and May 1986 the Earth will pass reasonably close to the comet's orbit. Bearing in mind that the parent comet will not be too far away, some enhanced activity may possibly occur at these times. Unfortunately, there are few historical observations of the Eta Aquarids and Orionids made at previous returns of the comet, to assist in making any predictions. As the present-day orbits of the Orionid and Eta Aquarid particles are not identical with that of the comet, it is possible no variation in shower activity will be observed. Nevertheless, world-wide groups of amateur meteor observers will be making visual meteor counts during the two showers and, indeed, it is likely that accurate Orionid and Eta Aquarid rates will be obtained over the period 1984 to 1987. This will provide important information on the distribution of meteoroid particles within the two streams probably linked with Comet Halley. Observations of the brightness distribution of the meteors seen will also provide information on the relative numbers of particles having different masses (and sizes) within the streams. During the coming apparition of the comet, the two important periods for meteor observation will be from October 16 to 26 1985, for the Orionids, and from April 29 to May 15 1986 for the Eta Aquarids. The latter of these showers occurs only three weeks after the comet's most favourable period of visibility.

With an object which arouses such great interest as Comet Halley, there was naturally much rivalry between the world's major observatories to be the first to recover the comet at the coming apparition. Up to the middle of 1985, its extreme faintness will mean that it is only accessible with the largest telescopes. It was widely maintained for some years that the comet would not be found before late 1983. However, this fact did not discourage professional astronomers from trying. On December 18 1981, a team from the California Institute of Technology made an unsuccessful search for the comet, using the 5.1-metre reflecting telescope. Their novel approach was to use a piece of electronic equipment called a Charge-Coupled Device (or CCD as it is known), placed at the prime-focus of the telescope. A CCD consists of a rectangular silicon wafer, 10 mm × 15 mm, on the surface of which over 16,000 picture elements or 'pixels' have been etched. Photons of light falling on to the silicon chip produce electrical charges within the silicon material, the quantity of charge produced being proportional to the intensity of the incident light. The charge accumulated in each pixel element (a measure of the light intensity which has fallen on that element) is transferred to a computer for storage and analysis. The CCD is amazingly sensitive, because it can look at an incredibly faint object for a long time, and steadily accumulate the effect of the light photons incident upon its surface. In this way, sufficient light may be gathered from the source to enable an image of the object to be generated by the computer, and later displayed on a screen for visual examination by an observer. Using such a device the CalTech team made 24 exposures of 5 minutes each, and 12 exposures of 1 minute 40 seconds each, of the area of the sky within which Halley's Comet was thought to lie. The field of view of the system used was much larger than the probable errors on the position of the comet. By summing a number of the exposures made, the workers were able to show that the comet's nucleus was fainter than magnitude 25. This means that if the albedo (reflectivity) of the nucleus is about 0.5 (comparable to the satellites of Saturn) then its diameter is less than 3 km, far smaller than was at one time thought.

Similar investigations to the one described above were also conducted at two other major observatories during December 1981. Both were unsuccessful in locating the comet. A team of French observers used a Lallemand electronographic camera at the prime focus of the 3.6-metre Canada-France-Hawaii Telescope in Hawaii, and failed to detect the comet at a magnitude of 25.2 ± 0.2, a very similar result to the CalTech team. At about the same time, an attempt was made to locate the comet from the Kitt Peak National Observatory in Arizona, using the 4-metre telescope there. The observers took 20 separate nine-minute exposures with a sensitive TV camera, and then summed the images, after allowing for the assumed motion of the comet. They concluded that the comet was undetectable, and fainter than magnitude 24.3.

Searches continued through 1982, both in the United States and at the Siding Spring Observatory in Australia. As we noted in our opening chapter, success finally came on October 16. Using a CCD at the prime focus of the great Palomar telescope, Jewitt and Danielson made five exposures each of 8 minutes' effective duration. 'Definite images near the expected position and having the expected motion of P/Halley were noted,' read the official announcement on International Astronomical Union Circular 3737, in which the comet was designated '1982i'. The comet was a mere 0.6″ of arc west in right ascension from the position predicted by Donald K. Yeomans, corresponding to an error in the time of perihelion of only 0.3 days. It was a triumph of mathematical investigation. At the time of its recovery, the comet was 11.04 AU from the Sun, just beyond the orbit of Saturn. The magnitude, 24.2, made it the faintest comet ever observed, and only one other (Comet Stearns, 1927V) has been seen at a greater distance from the Sun (11.53 AU). When first seen, the distance between Halley's Comet and the Earth was 10.93 astronomical units.

Before long confirmation was obtained at other observatories. The comet brightened slowly; on January 14 1983 R.M. West and H. Pederson, using a CCD on the Danish 1.5-metre reflector at La Silla in Chile, gave the magnitude as 23.5—slightly brighter than expected, which could be an indication that even at this tremendous distance from the Sun the comet was starting to show signs of activity. It was clear that fascinating times lay ahead, and it is also certain that nobody who lives through the decade of the 1980s will ever forget the welcome return of Halley's Comet.

CHAPTER 10

FRIEND OR FOE?

'This comet was so horrible, so frightful, and it produced such great terror that some died of fear and others fell sick. It appeared to be of extreme length, and was of the colour of blood. At the summit of it was seen the figure of a bent arm, holding in its hand a great sword as if about to strike. At the end of the point there were three stars. On both sides of the ray of this comet were seen a great number of axes, knives and blood-coloured swords, among which were a large number of hideous human faces, with beards and bristling hair.'

So wrote the French doctor Ambroise Paré in 1528. It gives a fairly good picture of the terror inspired by comets in those days, and the idea of comets as evil omens dates back for thousands of years. To be honest, there is considerable doubt about what Paré actually saw, because no brilliant comet was recorded by astronomers for some years before 1528, and Halley's Comet did not return until three years afterwards. It is even possible that the good doctor's account refers to a display of aurora rather than a comet. But there is no doubt that comets were taken to be messengers of disaster, and even Shakespeare wrote:

'When beggars die, there are no comets seen;
The heavens themselves blaze forth the death of princes.'

Comet terror was rife in Halley's day. For instance, the bright comet of 1681 caused such alarm in Switzerland that the town council of Baden issued a stern edict: 'All are to attend Mass and Sermon every Sunday and Feast Day, not leaving the church before the sermon or staying away without good reason; all must abstain from playing or dancing, whether at weddings or on other occasions; none must wear unseemly clothing, nor swear nor curse.' Presumably the Badenese were glad when the comet disappeared.

The possibilities of collisions with comets have always been admitted, and there have been several panics. One was due to a contemporary of Halley's, the Rev William Whiston, who believed that the comet of 1680 would eventually return, colliding with the Earth and causing immense damage. There was also a scare in 1832, when Baron Damoiseau published a paper showing that on October 29 of that year Biela's periodical comet would cross the orbit of the Earth—though he added that the Earth would be nowhere near that position at the time. And in our own age we have had Dr Immanuel Velikovsky, an eccentric Russian-born psychiatrist, who believed that the planet Venus used to be a comet (!) and that it made periodical close approaches to the Earth, producing phenomena such as the convenient drying-up of the Red Sea at the moment when the Israelites were waiting to cross. Velikovsky became something of a cult figure, and even today he still has his supporters. The trouble with his theories is that they bear no relation to true science, so that it is impossible to discuss them logically.

Yet are there any records of a cometary collision? Possibly. On June 30 1908 a cosmic missile landed in the Tunguska region of Siberia, blowing pine-trees flat over a wide area round the point of impact. Luckily the region was uninhabited, but had the missile hit a city the death-toll would have been colossal. No meteoritic fragments were found (though admittedly it was 1927 before a scientific expedition reached the site), and most authorities now believe that the object was part of the nucleus of a small comet, which was made up chiefly of ice and therefore evaporated on or just before landing. There is even a chance that the Siberian missile came from Encke's periodical comet.

If the nucleus of a large comet struck the Earth it would unquestionably do a great deal of damage, but the tails are harmless. The Earth seems to have passed through the tail of the Great Comet of 1861, and the only effect noted was a slight luminosity of the sky; a writer named E.J. Lowe reported that the comet itself looked hazy on the night of June 30, and that the local vicar had to light candles in the pulpit of his church—though whether this was genuinely due to the comet seems debatable. The next occasion brings us back to Halley's Comet, and again it is just possible that the Earth passed through the tail in 1910. Mary Proctor, a well-known British astronomer (daughter of the even more famous R.A. Proctor) wrote:

'In the forenoon of May 19 certain peculiarities observed suggested that our planet may have been actually immersed in the cometary débris of the train of Halley's Comet. These consisted of a peculiar iridescence and unnatural appearance of the clouds near the Sun, and a bar of prismatic colours on the clouds in the south. This, combined with the general effects of the sky and clouds—for the entire sky had a

Fig 10.1 *Devastation caused by the 1908 Siberean Meteor.*

most unnatural and wild look—would have attracted marked attention at any other time than when one was looking, on this occasion, for something out of the ordinary. According to the observations made by Professor Barnard at the Yerkes Observatory, the sky had been watched carefully during the forenoon of this date, but nothing unusual had appeared until close to noon, when the conditions became abnormal. Later on in June, and for at least a year afterward, slowly moving strips and masses of luminous haze were observed in the sky, which were not confined to any one part. Reports of unusual phenomena were received from the Transvaal, and from elsewhere in southern climes.'

It is worth noting that on May 18-19 it had been predicted that the comet would pass in transit across the face of the Sun. An American astronomer, Ellerman, went to Hawaii, where conditions were expected to be favourable, specially to find out whether any trace of the comet could be detected. Nothing was seen—an extra proof, if proof were needed, of the wraithlike nature of even a bright comet. However, it is also on record that an enterprising salesman in Dallas, Texas, made a considerable sum of money by selling what he termed comet pills—designed to ward off the evil influence of the comet, though he did not explain exactly how his pills were meant to work!

Rather surprisingly, there has been a recent revival of the old idea that comets may be harmful. This time the reasoning is not mystical or superstitious, but purely scientific, and has the backing of one of the world's great astronomers: Professor Sir Fred Hoyle. With his colleague Dr Chandra Wickramasinghe, Hoyle has put forward the unexpected theory that life on Earth was first brought here by a comet, and that comets may today deposit harmful substances in our atmosphere, causing epidemics ranging from influenza to bubonic plague.

The first glimmerings of such a theory go back to the work of Svante Arrhenius, a Swedish scientist whose work was outstanding enough to win him a Nobel Prize. Arrhenius pointed out, quite correctly, that we still have no definite knowledge of how life originated. (This is true even today.) According to Arrhenius' 'panspermia' theory, life was brought here by means of a meteorite, which landed in a suitable location and began the spread of living things all over the globe. The theory never proved popular, because it seemed to raise more difficulties than it solved, and certainly it is very hard to see how life could originate in a small body such as a meteorite.

Hoyle and Wickramasinghe also believe that life originated in space, though in almost all other respects their picture is different from Arrhenius'. Their main argument is that the appearance of living from non-living material involves a whole chain of events, each of which is itself improbable—so that the chances of life having appeared directly on the surface of a body no larger than the Earth are extremely slim. However, it is now known that many organic compounds exist in space, and Hoyle maintains that a vast area of space is needed for all these improbable events to occur one after the other. Therefore, everything depends upon a life-bearing object landing upon a world which is suited to it. If this is so, the Earth merely acted as a 'receiving station' for life which had been produced far away. And instead of a meteorite, Hoyle suggests that the likely 'carrier' was a comet. Comets, after all, contain suitable substances, and are ideal for transporting material from one part of the solar system to another. It is even possible that life began well beyond our own part of the galaxy, and that the Sun collected comets during its passage through an interstellar cloud. Moreover, if there are something like 100,000 million comets moving round the Sun at a distance of a light-year or so, their total mass may be equal to that of a giant planet such as Uranus or Neptune. Finally, it is suggested that comets also brought water here, so that they are responsible for our oceans.

Following up this theory, Hoyle goes on to suggest that comets may still be bringing harmful viruses and bacteria in our direction; and if these substances are deposited in the upper air, they will spread out and produce epidemics. In some ways this is a rather frightening idea. At the moment, the dread disease smallpox has been virtually wiped out. Could it be re-introduced by a passing comet?

In this sort of situation Halley's Comet is obviously of special significance, because it is the only large comet to return regularly in a period of less than several centuries. Hoyle and Wickramasinghe believe that it may have had profound effects over the past few thousands or even millions of years, and that these effects are still likely to make themselves felt every time there is a close approach. Presumably this would involve a minimum distance less than that at the 1986 return, but once again it is impossible to be sure.

Medical opinion in general does not agree with the Hoyle-Wickramasinghe theory, and it is fair to say that most astronomers, too, treat it with considerable reserve. A great deal of research remains to be done before any definite conclusions can be drawn. But at any rate it is interesting to find that some of the world's leading astronomers have come back to the age-old idea that comets, notably Halley's, may sometimes be unfriendly rather than benign.

CHAPTER 11

SPACE-CRAFT TO HALLEY'S COMET

Our knowledge of comets has grown immeasurably during the past few decades. We have a better idea of their composition and their nature, and we accept them as important members of the solar system, despite their lack of mass. But several outstanding problems remain. How were comets born? And perhaps above all, what are their nuclei like?

Remember, nobody has yet had a proper view of a cometary nucleus. In the case of Halley, we assume it to be mainly icy, with a diameter of a few kilometres; but we do not really know, and Earth-based studies give us very little hope of finding out. The only alternative is to go and see for ourselves—or, rather, send up equipment to carry out the surveys for us. And by now, a comet probe is a very practical possibility. If we can send our space-craft past the remote planet Saturn with almost perfect accuracy, it should be easy enough to contact a comet much closer to us. The first point to be settled is: which comet?

Obviously it would be advantageous to select a really large one, with a bright nucleus and preferably a long tail. The short-period comets, even the more conspicuous ones such as D'Arrest's, are puny by comparison. Moreover, these small comets have spent a long time in the inner part of the solar system, so that many of their volatiles have been evaporated and lost. Neither do they have impressive tails. But the brilliant visitors which adorn our skies every now and then, such as those of 1811 and 1843, cannot be predicted. Their periods are much too long. They are not likely to be detected until they are within the orbit of Saturn and more probably within that of Jupiter, which is not long enough for a probe to be made ready and launched. Therefore we must rely upon the one bright comet which returns regularly: Halley.

This was recognised years ago, and various plans were drawn up in the United States. In some ways Halley's Comet itself is unco-operative. True, its orbit is known with great precision, but its movement is retrograde, and this complicates matters. One suggestion involved launching a probe in the late 1970s and literally swinging it round Saturn in 1983, using Saturn's strong pull of gravity to force the probe into a retrograde path; it would then catch up the incoming comet, and move along with it, so that the encounter would last for weeks.

The most ambitious NASA plan was the Halley Intercept, which was to be launched in the summer of 1985 to encounter the comet eight months later, after perihelion, at which time the distance would be 1.1 AU from the Sun and 0.65 AU from the Earth. The probe would have carried a payload weighing 125 kg, and would, it was hoped, be capable of sending back close-range pictures just as those obtained from the two tiny Martian satellites, Phobos and Deimos, by Mariner 9 more than 15 years earlier. After the Halley encounter, Intercept was designed to go on to rendezvous with Encke's periodical comet and possibly even on to two more comets, Borrelly and Tempel 2, in 1988.

Alas for such hopes! Gradually the economic situation became less favourable; NASA funds were savagely cut back, and project after project was either abandoned or modified. One casualty was the Lunar Polar Orbiter, designed to be put into a closed path round the Moon to survey the polar regions, which are the only parts of the surface not properly covered by the Orbiters and Apollo space-ships. A further Mars probe was also abandoned. And as President Carter was succeeded by President Reagan, things became even more bleak from NASA's point of view. They had to abandon their cherished VOIR probe, which was to go to Venus and continue mapping the surface by radar. Project Galileo, the Jupiter orbiter and entry mission, was significantly modified, though it was so advanced in the planning stage that it was not actually cancelled. But probably the greatest blow was that the whole idea of sending a probe to Halley's Comet was wiped out at a stroke of the Pentagon pen.

The decision was heartbreaking for the astronomers and space scientists. The Moon and the planets are always with us, but Halley's Comet will not again be within range until the year 2061. NASA had to admit defeat. Even in the highly unlikely event of funds being made available at the last moment, no probe could be built and launched in time—and Halley's Comet will not wait.

The only crumb of comfort came with the approval of a probe to intercept the periodical comet Giacobini-Zinner in September 1985, but even this involved using a space-craft, IC3, which had been in orbit round the Earth since 1978. It had been designed to study the Sun, and in particular the solar wind before interaction with the Earth's magnetic field, so that it carried a full complement of instruments, and had a large fuel reserve. On June 10 1982 a series of manœuvres began, involving a series of 'swing-by' passages round the Moon. The first of these was timed for March 30 1983; on October 20 1983 there will be a very close swing-by, and on the following December 21 the probe is scheduled to pass by the Moon at only 150 km above the lunar surface, giving it sufficient energy to put it into a path which would take it close to Giacobini-Zinner, after which it will continue to monitor the solar wind. It may, therefore, be of some value in the Halley mission; but Giacobini-Zinner is very puny as compared with Halley, and the whole mission is nothing more than a pale reflection of the ambitious original programme.

Therefore, our main hopes for 1986 rest with three projects: Europe's Giotto, Russia's Vega, and Japan's Planet A. Giotto (there are no prizes for recognising the origin of that name!) was conceived in the 1970s by ESA, the European Space Agency. It is to be a fly-by, passing the comet at a relative velocity of 68 km per second and penetrating the coma, approaching the nucleus to within about 500 km. It is necessary to say 'about', because nobody is absolutely certain where the nucleus is. It may not be central in the coma, but in any case Giotto will be in serious danger as soon as it enters the main coma, and few of the planners dare to hope that it will emerge unscathed.

Moreover, the probe will remain within useful range of the coma for a mere four hours; at that time the distance from the Sun will be 0.89 AU (that is to say, about 133,000,000 km). There is only one Giotto vehicle, so that there can be no second chance, and everything must be ready for the probe to be dispatched on July 10

1985. The launch vehicle will be an Ariane rocket. Considerable problems with Ariane have been experienced in the past, but recent results have been more encouraging, and there is at least a reasonable hope that the launching will be equal to its task.

Targeting will have to be very precise. The error at launch will probably be of the order of 100,000 km, but it will be possible to make mid-course corrections, as has indeed been done with all the planetary missions, and here the experience with the Giacobini-Zinner probe will be helpful. By the time that Giotto draws close to Halley, we will know exactly how the space-craft is moving and exactly where the coma of the comet is, particularly as there should be invaluable information from the leading Russian Vega vehicle, which will reach Halley first. The corrections should be complete 32 hours before the time of the encounter on March 13 1986, and there is no reason to doubt that this part of the programme at least will be carried through satisfactorily.

Giotto is not a large vehicle. Its cylindrical body is 1.8 m in diameter, and the payload mass is no more than 53 kg, but it will carry elaborate equipment. There will be a telescope for imaging the inner coma and—we hope!—the nucleus, plus devices for providing information about the composition of the comet and the sizes of the particles in both coma and tail; there will also be a magnetometer for making detailed investigations of the plasma. The cylindrical body will be covered with solar cells, and there will, of course, be ample provision for transmitting the information back to Earth and any space-stations which are in orbit at the time of the encounter (perhaps even the Space Telescope, scheduled for launching in late 1985). Giotto will be spinning in a controlled manner, making 15 revolutions per minute. If all goes well, the pictures received should be on a large scale; when Giotto is at its closest to the centre of the comet, at 500 km or so, the resolution should be down to about 20 m.

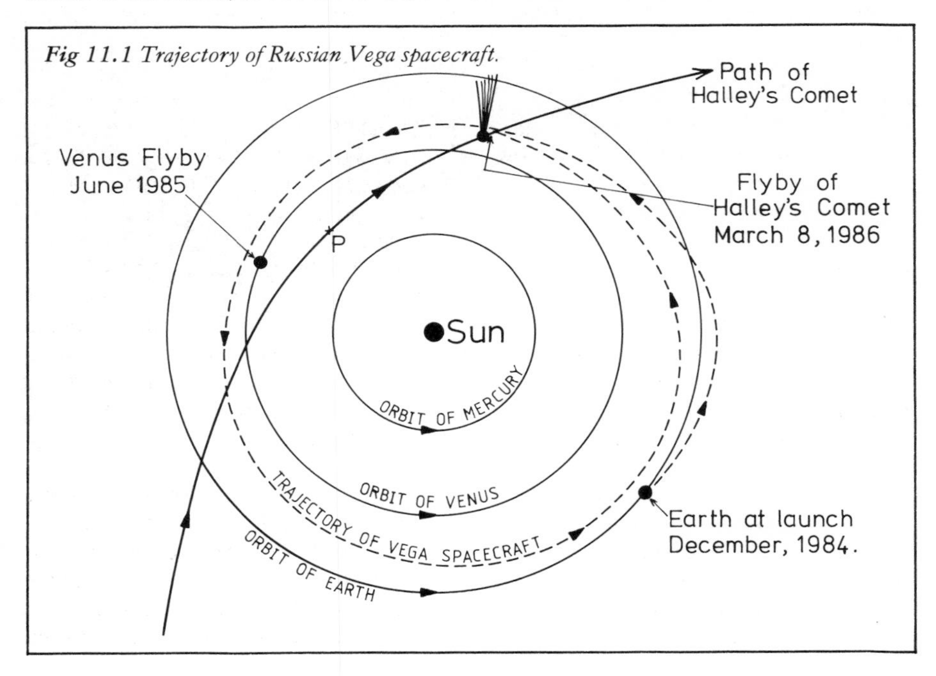

Fig 11.1 Trajectory of Russian Vega spacecraft.

It is painfully clear that collisions with particles will be a major hazard. Our knowledge of the sizes of particles in the coma, and in particular near the nucleus, is very scanty, and if Giotto is hit by an object the size of, say, a football there can be only one result. The only precautions which can be taken involve the addition of a dust-shield, multi-layered and gold-plated. It is hoped that the outer layer of the shield will break up any impacting particle, so that only fragments of it will reach the inner shield. With luck, this arrangement should prove adequate up to a late stage in the mission, but all information will be sent direct back to the Earth or a space-station, and no data will be stored on board the probe for transmission later. Indeed, there may not be a 'later'. Few of the planners expect Giotto to pass so close to the nucleus and survive.

What kinds of pictures can be expected? If all goes well, it is hoped to be able to measure the size and also the rotation of Halley's nucleus. Of course, the tail will also come under surveillance. In particular, we want to know whether the rate of evaporation is steady, or whether there are a few active emission areas on the crusty or dust-laden mass of the coma and nucleus. In view of the marked activity in the tail during the returns of 1835 and 1910 it is probable that the emission rate will be very irregular, with bursting and jetting, but one never knows. Even Halley's Comet is unpredictable in its behaviour.

Finally, Giotto will send back data about space conditions near the comet. There may well be magnetic fluctuations, and both cosmic ray and solar wind studies will be important.

Such is Project Giotto. It is ambitious, and there can be no rehearsal; the planners would have liked Halley's Comet to have come back not in 1986, but some years later, when space-research methods will have been refined and the economic outlook may be brighter. But at least there are grounds for optimism, and by the time that the comet draws back into the further reaches of the solar system it may have told us a great deal.

Next there is the complex Vega mission, planned by the Russians with co-operation from French and East European scientists. There will be two space-craft, each of which will encounter Halley's Comet—going there by way of Venus, and dropping capsules and even balloons into the atmosphere of that decidedly peculiar and hostile planet.

The idea of multi-target probes was suggested long ago, and have proved to be quite sound. Mariner 10 went by Venus en route for Mercury; Pioneer 11 encountered first Jupiter and then Saturn; Voyager 1 did the same, while Voyager 2 passed by both Jupiter and Saturn before being put into the path which will, it is hoped, lead to a rendezvous with Uranus in January 1986 and Neptune in August 1989. A vehicle of this type is aimed so that it will swing past its first target, and use the gravitational force of that body to send it on to the next. Of course, everything depends upon the relevant targets being in the right places at the right times, as the four giant planets were in the late 1970s; with Halley's Comet, Venus is similarly obliging.

In fact, the Russians have already scored great successes with their Venus programmes (though, strangely, they have had almost no luck with Mars, where the problems would be expected to be much less serious). Several Soviet vehicles have made controlled landings on Venus, and have obtained the only pictures direct from the surface; the last two, Veneras 13 and 14, did so in 1982, providing impressive views of the gloomy, rock-strewn landscape. Beautiful though it may look, shining down from the eastern or western sky almost like a small lamp, Venus is far from welcoming. The surface temperature is of the order of 500° Centigrade; the atmospheric pressure is 90 times that of the Earth's air at sea-level; the atmosphere itself consists mainly of carbon dioxide, and the clouds contain large quantities of sulphuric acid. Yet Venus is a fascinating world, and it has much to tell us. In the early period of the solar system, when the Sun was less luminous than it is now, Venus may have had oceans, and there may even have been primitive life.

According to the detailed information given by the Russians at the Congress of the International Astronomical Union held in Greece in August 1982, the first Vega will be launched on December 15 1984. The flight to Venus will take between 174 and 176 days, so that the encounter will take place in June 1985. Two days before closest approach, the Venus lander and atmospheric probes will be released, to finish their journeys independently. During the descent towards the surface, the instruments will send back data about the temperatures, pressures and chemistry of the atmosphere; they will make gentle landings, controlled partly by parachutes; once on the surface they will scoop up samples of soil, analyse them, and send back the results, as NASA's Viking probes did when they reached Mars. The drop through Venus' atmosphere will take ten minutes, and before the final manœuvres the landers will be chilled, so as to help in withstanding the tremendous heat in the lower atmosphere. The balloons will simply float about, transmitting information from various levels. Meanwhile, the main probe will be on its way to a rendezvous with Halley on March 8 1986, about a month after the comet has passed perihelion.

The minimum distance between the probe and the comet will be 10,000 km, with a relative velocity of 78 km per second, somewhat higher than Giotto's. The passage through the coma will take place between March 6 and 12, and there will be five separate stages of data-finding, from two days before closest approach to two days afterwards—provided, of course, that Vega survives. The vehicle will be spin-stabilised, and, as with the other probes, there will be dust-shields to provide as much protection as possible. There will be a television camera, with one narrow-angle and one wide-angle lens, mounted upon a scan platform now being constructed in Czechoslovakia. Emissions from the coma and nucleus will be studied by means of an infra-red spectrometer, mounted alongside the television camera on the scan platform; there will be a magnetometer, plus a high-energy particle detector and devices for studying the plasma in the immediate region of the comet. Particles of from 10^{-10} to 10^{-18} grammes in mass will be investigated.

Again we come back to that all-important question: where exactly is the nucleus? Results already obtained indicate that the centre of light may be about 1,000 km ahead of the nucleus on the sunward side of the comet, though it is impossible to be sure. The Russians hope that they will be able to give more accurate estimates about 48 hours before encounter, so that they can relay the information to help in re-targeting Giotto if necessary. With luck, Vega will also be able to obtain a time-exposure picture of the nucleus, something which will be much easier with a stabilised space-craft than with the spinning Giotto. By that time it is expected that the precise position of Vega will be known to within 100 km.

The second Vega, a twin of the first, will follow shortly afterwards. If Vega 1 is destroyed at 10,000 km from the comet, then Vega 2 will be sent by at greater range; but if Vega 1 survives, its twin will be commanded to pass within 5,000 km—no closer, because it would then be impossible to obtain detailed results owing to the rapid velocity of the probe relative to the comet.

Meanwhile, the Japanese have been working upon what may be termed a cut-price programme, again involving two probes; a preliminary mission, known by its type number of MS-T5, and Planet A itself. Both are much smaller than Giotto, but the Japanese genius for building miniaturised equipment may go a long way towards offsetting the weight problem. It is true that the Japanese have not yet attempted a major mission in space, but they have dispatched a dozen satellites into orbit since their first foray in 1970, and the Halley probes will have payload masses of over 100 kg.

The first, the test mission MS-T5, will be sent up by a solid-fuel launcher as early as January 1985, and will carry out detailed investigations of the solar wind as well as its main task. By November it will be about 15,000,000 km from the Earth, and the Japanese will have gained vital experience in controlling it—experience which will be used to improve the performance of Planet A, which will be sent up in August 1985. (The exact date has not yet been decided; the so-called 'launch window' lasts for a week.) The fly-by error is not likely to exceed 20,000 km. It is hoped to image the coma and even the nucleus in ultra-violet, as well as carrying out other experiments basically similar to those on Giotto and the Vegas. The Japanese are leaving nothing to chance, and are even constructing a special 64-metre antenna so that they can keep track of both their probes.

All these missions will be closely linked with Earth-based observational programmes, and, if all goes well, with the Space Telescope. Space research is now an integral part of astronomy itself, and Halley's Comet provides us with an unrivalled opportunity for solving some fundamental problems. It is a pity that the return comes so soon instead of in the 1990s, but we must not complain. If the return had taken place 20 or 30 years ago, we would have been unable to take advantage of it, and we would have had a long wait.

CHAPTER 12

SUMMARY AND APPENDICES

Previous returns

Examinations of Chinese historical records of so-called 'guest-stars' have revealed some interesting ancient observations of Halley's Comet. In particular, modern high-speed computers have been able to integrate the orbit of the comet backwards in time, making due allowance for the perturbations of all major planets and the effect of the non-gravitational force. The work of Donald Yeomans and Tao Kiang has been of particular importance in extending our knowledge of the motion of Comet Halley back to 1404 BC. The essence of much of these efforts involves the determination of the times of perihelion passage from ancient Chinese records and a comparison of these with the computations. More recently, Joseph Brady has carried out a series of calculations in which he determined times of perihelion passage for Comet Halley as far back as 2647 BC. Many of his dates for the returns before 87 BC disagree with those obtained by Yeomans and Kiang.

Throughout this section, the return of 1910 is referred to as 'Return-1', the previous one in 1835 as 'Return-2' and so forth. The time of perihelion passage is denoted by the letter T, and the symbol Δ denotes the distance of the comet from the Earth. The dates are from the results of Yeomans and Kiang.

Return	Year	Perihelion passage	Remarks
-41	1059 BC	T = Dec 3	Earliest probable recorded observation of Halley's Comet.
-29	240 BC	T = May 25	This is the earliest return about which there is something more than a mere mention in Chinese annals.
-28	164 BC	T = Nov 12	Curiously, this rather favourable apparition (minimum Δ = 0.1 AU) went unobserved in Sept/Oct of 164 BC.

Data are available for all returns of Halley's Comet after that of 164 BC.

-27	87 BC	T = Aug 6	
-26	12 BC	T = Oct 10	Observed: Aug 26 to Oct 20. Contrary to a popular myth, Halley's Comet certainly was *not* the 'Star of Bethlehem'.

Return	Year	Perihelion passage	Remarks
-25	66 AD	T = Jan 25	Observed: Jan 31 to Apr 11. Was Halley's Comet, described as like 'a sword hanging in the sky', the sign in the heavens which foretold the destruction of Jerusalem?
-24	141 AD	T = Mar 22	Observed: Mar 26 to May ? Moderately close approach of comet to Earth on April 22 (minimum Δ = 0.17 AU).
-23	218 AD	T = May 17	Observed: Apr to May.
-22	295 AD	T = Apr 20	Observed in May.
-21	374 AD	T = Feb 16	Observed: Mar 3 to May ? An Earth-comet close approach took place on Apr 2 (minimum Δ = 0.09 AU).
-20	451 AD	T = Jun 28	Observed: Jun 10 to Aug 16.
-19	530 AD	T = Sept 27	Observed: Aug 28 to Sept 27.
-18	607 AD	T = Mar 15	Observed: Apr 18 to July ? An Earth-comet close approach took place on Apr 19 (minimum Δ = 0.09 AU).
-17	684 AD	T = Oct 2	Observed: Sept 6 to Oct 24. Earliest recorded drawing of the comet in 684 AD appeared in the Nürnberg Chronicles, published in 1493.
-16	760 AD	T = May 20	Observed: May 16 to July ?
-15	837 AD	T = Feb 28	Observed: Mar 22 to Apr 28. Most spectacular return ever of Halley's Comet. The comet passed within only 0.04 AU of the Earth on Apr 11, 42 days after perihelion. The apparent magnitude was -3.5, comparable with the planet Venus, and the tail stretched 93° across the sky.
-14	912 AD	T = July 18	Observed: July 19 to July 28.
-13	989 AD	T = Sept 5	Observed: Aug 11 to Sept 11.
-12	1066 AD	T = Mar 20	Observed: Apr 1 to Jun 7. This apparition of Halley's Comet was immortalised on the Bayeaux Tapestry, woven to commemorate the Battle of Hastings, which occurred later in the same year. In mid-Apr, when quite near the Earth, the comet's brightness was said to rival the planet Venus, at probably about apparent magnitude -3.
-11	1145 AD	T = Apr 18	Observed: Apr 26 to July 9.
-10	1222 AD	T = Sept 28	Observed: Sept 3 to Oct 23.
-9	1301 AD	T = Oct 25	Observed: Sept 15 to Oct 31. Seen by Giotto di Bondone (1267–1337) the Florentine artist who painted 'Adoration of the Magi'.
-8	1378 AD	T = Nov 10	Observed: Sept 26 to Nov 10.
-7	1456 AD	T = Jun 9	Observed: May 26 to July 8. Halley's Comet condemned by the Catholic Church and Pope Calixtus III as 'an agent of the Devil'.

Return	Year	Perihelion passage	Remarks
-6	1531 AD	T = Aug 26	Observed: Aug 1 to Sept 8.
-5	1607 AD	T = Oct 27	Observed: Sept 21 to Oct 26.
-4	1682 AD	T = Sept 15	Observed: Aug 24 to Sept 22. Seen by Edmond Halley himself. He decided that this comet was the same as that of 1607 and 1531. He predicted it would return in 1758.
-3	1759 AD	T = Mar 13	Observed: 1758 Dec 25 to 1759 June 22. The first 'predicted' return.
-2	1835 AD	T = Nov 16	Observed: 1835 Aug 5 to 1836 May 19. An Earth-comet close approach took place on Oct 10 (minimum $\Delta = 0.05$ AU).
-1	1910 AD	T = Apr 20	Observed: 1909 Aug 25 to 1910 Jun 16. The first return in 'photographic times'. A transit of the comet across the Sun occurred on May 18. The comet approached Earth to within 0.14 AU shortly afterwards.

Brightness

Conventionally, the total apparent magnitude, M_1, of a comet is expressed by the equation:

$$M_1 = M_0 + 5\log\Delta + n\log r$$

where Δ is the comet's distance from the Earth and r is the comet's distance from the Sun (both are measured in Astronomical Units). M_0 is called the total absolute magnitude of the comet and $M_1 = M_0$ when $\Delta = r = 1.0$ AU.

The value of the quantity n is different for different comets, and even for a given comet may vary (ie, before and after perihelion).

From an analysis of the 1910 brightness estimates of Halley's Comet, the pre-perihelion total magnitude could be represented by the formula:

$$M_1 = 5.0 + 5\log\Delta + 13.1\log r$$

The brightness after perihelion showed an interesting dip between $r = 0.6$ and 1.0 AU, before the characteristic post-perihelion brightening occurred. Predictions of the post-perihelion brightness of Halley's Comet are made by fitting empirical curves to the 1910 magnitude estimates.

Another analysis was carried out on brightness estimates which were described as nuclear magnitudes. These nuclear magnitude estimates M_2, were made using large telescopes for great distances of the comet from the Sun. These data could be fitted by the formula:

$$M_2 = 7.5 + 5\log\Delta + 10\log r$$

This brightness is not true nucleus brightness, but refers to the apparent or photometric nucleus of the comet.

Orbital elements

A drawing showing the orientation of the orbit of Comet Halley in space has already been given in Fig 9.2. Six parameters are used to define uniquely the position and path of the comet in its orbit. The shape and size of the orbit for an ellipse are defined by the eccentricity, e, and by the semi-major axis, a. The orientation of this orbit in space is specified by a further three parameters. These are the inclination, i, of the orbital plane to the plane of the ecliptic, the longitude of the ascending node, Ω (which is the angular distance from the ascending node to a specific direction in space known as the vernal equinox, ϒ) and thirdly, the argument of perihelion, ω, which is the angular distance between the perihelion point and the ascending node measured in the direction of the comet's motion. The sixth orbital element is the time of perihelion passage of the comet denoted by the letter, T, ie, the date on which it passes the perihelion point.

For any celestial body the orbital elements do not remain constant. Perturbing forces due to other bodies in the solar system cause the orbital elements to slowly change with time. Therefore, for any object the calculated osculating elements are only valid at a particular instant in time known as the epoch of the orbit, although for low precision computations they may be used for a few months before and after the epoch.

For the 1986 return of Halley's Comet, the set of osculating orbital elements are based upon the work of Donald Yeomans, and are given as follows:

Epoch	1986 Feb 19.0 (ET)	
Perihelion passage	1986 Feb 9.44394 (ET)	
Orbital eccentricity	0.9672759	
Semi-major axis	17.94104 AU	
Orbital inclination	162°.23930 (retrograde)	1950.0
Longitude of ascending node	58°.14538	1950.0
Argument of perihelion	111°.84804	1950.0

Additional information

Perihelion distance	0.5871045 AU
Aphelion distance	35.294976 AU
Heliocentric distance of ascending node	1.81 AU
Heliocentric distance of descending node	0.85 AU
Distance of perihelion above ecliptic plane	0.17 AU
Distance of aphelion below ecliptic plane	9.99 AU

The above orbital elements were derived from 625 observations over the interval 1835 Aug 21 to 1982 Dec 10.

Observation

For those not interested in making useful scientific observations of the comet, requirements in terms of equipment are small. Any comet is a diffuse, rather indistinct object and hence is best observed with a pair of good binoculars (if brighter than about eighth magnitude), or a wide-field, short focal length telescope in conjunction with a low-power eyepiece. Even a 75 mm telescope with an eyepiece giving a power of ×25 will show comets as faint as ninth magnitude. Of some importance is the choice of a good observing site, preferably away from city lights and pollution, with a reasonably unobstructed horizon in the direction required. When the comet is faint and is being observed in a dark sky, make sure to let your eyes get well dark-adapted. Do not ruin your night vision with a bright torch, but use instead a red cycle lamp, suitably dimmed with layers of paper.

Of course, many observers will try to photograph the comet and this is advantageous in that it provides a permanent record of its appearance at that time. Always remember to note the date, time and length of exposure of any picture attempted. Very good cometary photographs may be obtained with lenses of modest focal length, say from 300 mm to 500 mm, and working at f/5.6 or better. When the comet is at its brightest, it should be possible to obtain reasonable results using a fixed camera, provided fast film is used, the exposure is kept short, and the sky background is fairly dark. Hard and fast rules are difficult to give, but after a few trial runs even a complete beginner should achieve some success. One problem when longer exposures are attempted is that the comet is in motion relative to the background stars, and also the stars themselves are moving from east to west. If an exposure of even a few minutes is contemplated, some form of guiding will be essential. This is usually achieved by having a small guide-telescope, fitted with cross-wires, fixed to the photographic instrument. A rigid mount and some form of hand-operated slow-motion drive will be required, to allow the operator to follow the movement of the comet by keeping its image centred on the cross-wires in the guide-telescope. When attempting photography in a twilight sky, fogging of the film may be a nuisance, especially as the comet is always close to the Sun when at its most spectacular.

For those interested in making some form of scientific contribution to the study of the comet, there are many opportunities for interesting and valuable work. In the case of a highly active comet such as Halley, the visual appearance of the coma and tail structure is subject to continual change, often on a time scale of only a few hours. For this reason, world-wide coverage by teams of astronomers is essential. For those unwilling or unable to photograph the comet, careful drawings may be made to show the important structural features, and in particular, any fine detail present in the head or tails. Attention should be given to the degree of condensation in the centre of the coma, its diameter and any irregularities in shape or brightness. The presence of any asymmetry in the coma, especially the appearance of jets or haloes, should be carefully noted. Start the drawing, using a low power eye-piece, with a higher power to fill in any fine detail, if present. A crisp, bright image under low power is highly preferable to a large faint and indistinct image under high magnification.

It is important to be able to estimate angular distance in the telescopic field in order that the size of the coma and position of the comet relative to nearby stars seen in the telescope may be determined. The angular diameter of the field of each eye-piece should be known. It is also useful to know the orientation of the telescopic field. The direction of the east-west line may be found by allowing a star to drift across the field. Once this is known, the position angles of any tail features, jets, streams or elongations of the coma may be ascertained. The position angle is measured eastwards from the north point. South is normally uppermost in the field of an astronomical telescope, but the exact orientation needs to be found.

Another useful field of cometary study is the assessment of the brightness of the comet as a function of its changing distance from the Earth and the Sun. Estimates of cometary brightness (magnitude) are notoriously unreliable due to the fact that a comet is an extended light source, not a point. An observation of the total brightness of a comet, as well as a separate one for the central condensation, is made by comparing the comet with nearby stars of about the same brightness, put out of focus so as to represent the comet as closely as possible. Magnitude estimates should be made with a low-power eyepiece, and a sketch of the comet and comparison stars used should be made to enable the stars to be identified later. The observer should also state the date and time of observation, the seeing conditions, and show the orientation of the telescopic field on his sketch.

For those who are keen to participate in the serious observation of comets, it is important to contact and have correspondence with other astronomers who are undertaking this type of work, and in particular those people who have considerable experience in this field. Most countries have national astronomical associations, with sections specialising in cometary study. In addition, many towns and cities have their own local astronomical societies and these should be contacted for helpful advice. They might even have some telescopes which you could use for your observations. In any case, a shared interest in any topic does much to increase its pleasure and usefulness. A full observational programme will also be carried out by the Comet Section of the British Astronomical Association. Full details may be obtained from the Association at Burlington House, Piccadilly, London W1V 0NL.

The International Halley Watch

The idea of an International Halley Watch (IHW) was formulated by Louis Friedman, then of the Jet Propulsion Laboratory at Pasadena (California), in 1979. The principle was to set up an organising body which would co-ordinate world-wide ground-based observations of the comet throughout the 1985-86 apparition. In late 1980, Ray Newburn became Acting IHW leader, and the Jet Propulsion Laboratory became the Lead Centre for United States activity. The idea was soon supported by NASA and by many United States scientific boards. In 1981, NASA set up an International Steering Committee of international scientists to organise the IHW programme. In November 1981, the Steering Committee selected a group of cometary specialists to become Discipline Specialists in seven specific technical areas of research. This organisation was enthusiastically accepted by the Executive Council of the International Astronomical Union, the controlling body of world astronomy, in August 1981.

Since 1981 the aims and objectives of the IHW have been firmly established. The organisation is controlled from two main offices, one in Pasadena, under Ray Newburn, and the other in Bamberg (West Germany) under the direction of Jürgen Rahe, a senior member of the IHW. Financial support for the two offices is being provided by the respective governments of the two countries. It has been recognised by the Halley space missions—Giotto, Vega and Planet A—that their chances of success will be significantly increased by mutual co-operation with the IHW team. The IHW will supply the missions with ephemeris data, and will place the brief space-craft encounters with Halley's Comet into the whole context of the 1985-6 return of the comet.

Four main objectives of the IHW have been given, as follows:

1 To encourage and support any scientifically valid means of studying Halley's Comet.

2 To co-ordinate the activities of ground-based observers with the space missions, so as to obtain the best possible results from the whole body of observations.

3 To set useful standards for observations in each of the seven scientific disciplines, although there is no obligation that all who participate in the IHW must meet these standards.

4 All properly documented data (that is to say reduced data, not interpretations) to

be published in a Halley Archive in 1989. This Archive will complement the usual scientific papers to be found in astronomical literature.

The seven scientific disciplines co-ordinated by the IHW are listed below, with the names of the co-ordinators:
Astrometry (D.K. Yeomans, R.M. West).
Infra-red Spectroscopy and Radiometry (R. Knacke, T. Encrenaz).
Large-Scale Phenomena (J. Brandt, J. Rahe, M.B. Niedner).
Near-Nucleus Studies (J. Rahe, Z. Sekanina, S. Larson).
Photometry and Polarimetry (M.F. A'Hearn, V. Vanysek).
Radio Science (W.M. Irvine, F.P. Schloerb, E. Gerard).
Spectroscopy and Spectrophotometry (S. Wyckoff, P. Nehinger, M.C. Easton).

Three further areas of great importance to the success of the IHW have been recognised. They are:

The Amateur Observation Net (co-ordinator, S.J. Edberg).
This has been organised to encourage useful participation by amateur astronomers in the IHW. Amateur enthusiasm for Halley's Comet is bound to be intense, and the aim of the Net is to direct this energy so as to supply useful scientific data. Five areas of study have been given for amateur participation:
(1) visual observations of the comet
(2) photometry
(3) spectroscopic observations
(4) photoelectric photometry
(5) observations of the Eta Aquarid and Orionid meteor streams.

The IHW News-Letter (editor, Stephen J. Edberg)
This publication will be produced either on an irregular or a quarterly basis. The hope is to keep open the lines of communication between the IHW Lead Centre, the Discipline Specialists, observers, space-missions and all other interested parties. The publication is prepared by the Jet Propulsion Laboratory, under a contract with NASA.

The Halley Archive (editor, Zdenek Sekanina)
This will not be a collection of interpretations or of ordinary papers, though an index to published papers may well be included. The Archive will consist of reduced data, arranged chronologically. Since quantitative imaging data and some spectra cannot easily be presented in book form, these may be included on a video disk. The Archive will be available for the use of present and future scientists in deriving a clearer picture of cometary phenomena.

INDEX

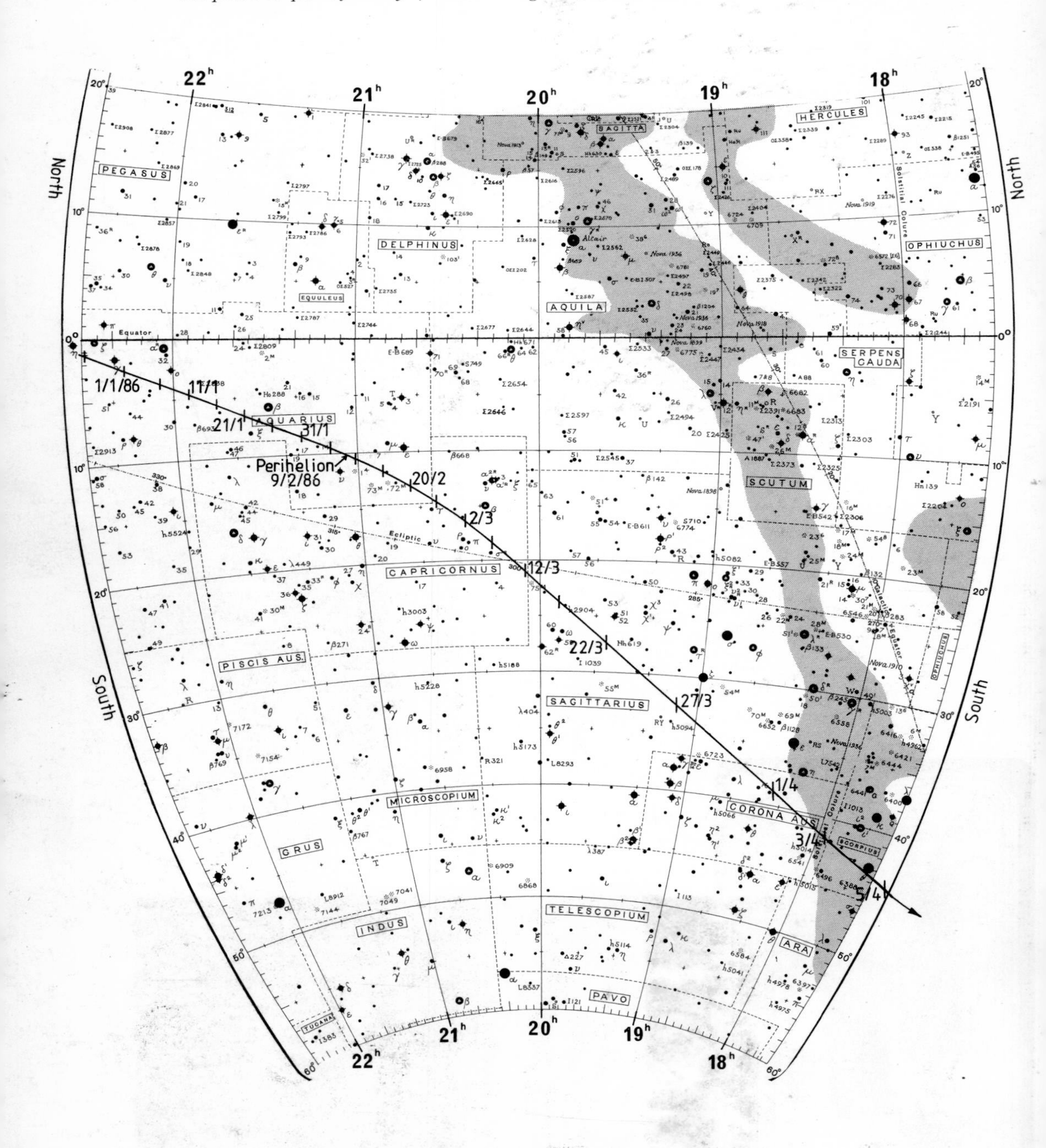

The predicted path of Halley's Comet during its 1985–6 return.